AF262981

NOTICING

Books by Richard Louv

Noticing
Our Wild Calling
Vitamin N
The Nature Principle
Last Child in the Woods
Fly-Fishing for Sharks
The Web of Life
FatherLove
101 Things You Can Do for Our Children's Future
Childhood's Future
America II

NOTICING

*Intimate Encounters
with the Natural World*

RICHARD LOUV

Photographs by the Author

ALGONQUIN BOOKS

LITTLE, BROWN AND COMPANY

Algonquin Books of Chapel Hill / Little, Brown and Company
Hachette Book Group
1290 Avenue of the Americas, New York, NY 10104
algonquinbooks.com

First Edition: May 2026

Algonquin Books of Chapel Hill is an imprint of Little, Brown and Company, a division of Hachette Book Group, Inc. The Algonquin Books name and logo are trademarks of Hachette Book Group, Inc.

The publisher is not responsible for websites (or their content) that are not owned by the publisher.

The Hachette Speakers Bureau provides a wide range of authors for speaking events. To find out more, go to hachettespeakersbureau.com or email hachettespeakers@hbgusa.com.

Contact your local bookseller or special.markets@hbgusa.com regarding special discounts for bulk purchases.

Photographs by the author

ISBN 9781643753034
Library of Congress Control Number: 2025950658

Printing 1, 2026

LSC-C

Printed in the United States of America

*To Kathy Frederick Louv
and the memory of my parents*

I respectfully acknowledge that I reside in the unceded
territories of the Ipai-Tipai Kumeyaay, the Payómkawichum,
the Kuupiaxchem, and the Cahuilla.

I believe that much unseen is also here.

— WALT WHITMAN

CONTENTS

NOTICING

INTRODUCTION

*I arise in the morning torn between a desire to improve . . .
the world and a desire to enjoy . . . the world.
This makes it hard to plan the day.*

— E. B. WHITE

NOTICING IS THE beginning of wonder.

Wonder, as the memoirist Valarie Kaur defines it, is the act of looking at something in nature, including people, and thinking, "You are a part of me I do not yet know."

Deep noticing represents an accumulation of old, new, and still-developing skills — sensory, scientific, artistic, and spiritual. It is also about confronting despair and finding hope in a time when all our sanity seems to be wearing thin.

Not long after my wife, Kathy, and I moved to the mountains, I joined a team to monitor wildlife trail cameras. These trail cams, as they're usually called, are stationary battery-powered infrared cameras housed in metal cases and usually mounted on trees. Their shutters are triggered by movement, and they can capture still photos or videos, day or night.

My trail-cam partner is Judy Kemp, a retired physician.

Every few weeks, Judy and I drive to Volcan Mountain — or Hahachepahg, its Kumeyaay name, which means "where the water comes from."

Today Judy is driving, so I hop out to unlock the gate to the 31,000-acre Volcan Mountain Wilderness Preserve. In her Jeep, we wind upward through rising hillsides spiked with burned and leaning trees. At an elevation of almost five thousand feet, we park beneath a grove of reaching ponderosas and cedars. Then we walk along a trail of switchbacks flanked by poison oak.

At a high turn on the path, we find the motion-activated night-vision camera in its locked box strapped to a tree. We collect the camera's memory card, replace the batteries, and insert a new card. Then we return to the car and drive down to the Volcan Mountain Foundation's education center. There, hunched over a laptop, we're eager to see what pops up on the screen. Maybe we'll find another parade of foxes or coyotes, perhaps bobcats. Or gentle opossums — with their opposable thumbs, prehensile tails, fainting spells, and three-year lifespans. Also some California jays and mountain doves, or an intense owl staring at the lens. Occasionally, a mountain lion (or cougar or puma — pick a name)[i] — perhaps a mother with two grown cubs — will make an appearance. We enter the data and send the photos on their way. These will be considered later by researchers studying Volcan's evolving ecology.

On today's trip up the slope, after an unexpected snow and windstorm, large sheets of ice and snow cover much of the road and hiking trail. Fallen trees are on our path. On foot, we crouch and weave our way through broken branches. A mountain lion has crossed a hard patch of snow.

I reach down and touch the impression of the big cat's track.

Later, we sit on cut logs in a dry, sheltered grove of pines and rest. Purple mushrooms surround us. Judy says she's been thinking about her perceptions of nature, how she notices the visual details, mainly when

---

i   Mountain lions, cougars, pumas, and catamounts are all the same species. Ranging from Chile to the Canadian Yukon, they also go by panther, red tiger, screamer, and thirty-three other names. Adult males usually weigh up to 150 pounds and, nose to tail, can be more than eight feet long. Females are usually a foot shorter.

birding. "But lately," she adds, "I've been wondering if I'm missing something. It's a feeling."

In the past, she avoided what she considered overanalyzing nature, believing it would lessen the experience and fill it with words. But now, she wonders, "Is there *something*… something I'm not seeing or feeling?"

Her voice trails off. She looks down. Her eyes are searching.

THE IDEA FOR THIS BOOK came to me one morning during the later stages of the COVID-19 pandemic.

At that time, most face-to-face interviews, in-person conversations, and air travel were still replaced by phone calls, email, and Zoom — mixed blessings.

I was walking up our hill, feeling, well, disembodied, disconnected from people and activity and my usual life. At that moment, I realized how absorbed I was with the chatter and combat of acorn woodpeckers, the individual characters of clouds, the scent of oaks and pines and cedars, their limbs lifting in the wind, and the leaf-litter passageways that led into darkness or light.

A group of turkeys, two females and a gaggle of chicks, shot across the road in front of me. One chick held back. I stopped to wait for it to catch up with the rest. Watching them, I was curious about how they notice life in their world.

From the writings of naturalist Joe Hutto, I knew that turkeys come genetically preloaded with survival skills. They arrive in this world already knowing the difference between a rattlesnake and a garter snake. They understand that the garter snake means them no harm. But throughout their lives, they go beyond their survival manual; they learn. Having never seen a garter snake in the flesh, a turkey will stop and blurt a particular sound, and the other turkeys will come running, and they study the snake. Like a model in an art class, the snake will allow itself to be observed.[1]

And when the turkeys *know* that snake, they move on, pecking their way through turkey territory. They see each field, each tree, each gully, each animal. From these details, they construct a world in their minds.

As I watch them gather their seeds of knowledge, I was sure I could detect their joy of discovery.

And I recalled how, during my decades as a reporter and columnist, I would enter a room and begin to gather the details — the clock on the wall, the pattern and texture of the wallpaper, the toy on the floor. When the people I came to interview walked into the room, I peeked at their shoes, their clothes, their eyeglasses. Then, in the surrounding neighborhood, I gathered more details and later constructed a hologram of that person or place.

Sharing the finished piece of writing would produce a sense of satisfaction. But the product was not where the joy was. The joy was in the *noticing.*

I felt that as a boy, during countless hours spent in the woods, when everything seemed new or reborn. I want to feel that joy again.

In 2005, my book *Last Child in the Woods* introduced the term *nature-deficit disorder.* The term isn't a formal medical diagnosis, though it might deserve to be. What it provides, however, is a doorway into a larger conversation. While researching the book, I could identify only about sixty studies worldwide on the benefits of nature experiences. We've since seen some progress. Today, over fifteen hundred scientific studies have underscored the importance of time spent in the natural world to the cognitive, physical, and emotional health of children and adults.[2]

Indeed, an international movement to connect families and communities to nature — and to each other — has emerged, and it's producing tangible results. Pediatricians in the United States and Canada have begun to prescribe time spent in nature. The number of nature-based schools is skyrocketing. Urban planners and political leaders have taken on the cause.

Even as our awareness grows, the divide widens. In 2025, research in the United States, Canada, and the United Kingdom reported that in Britain, two-thirds of Generation Z (born between 1997 and 2012) say they don't venture outside "for days at a time," as the *Independent* reported. The same is true for more than half of millennials. Generation Alpha (2010 to 2024) appears to be following a similar trend.[3] If alienation from the natural world continues at this pace, the health of both people and the planet will be threatened faster than we may have imagined. So, how do we find our way beyond this moment?

FOR SEVERAL YEARS I've been dealing with my own nature deficit by immersing myself in the mountains east of San Diego, California.

In the following pages, I'll share some of my experiences. I've also been asking folks here about their often-unconscious acts of what I call *deep noticing*.

You'll meet a nature therapist who helps people rediscover their hidden or lost selves in a mountain forest; a man who, for years, has driven sixty miles weekly to plant pine seedlings on the same mountain, which has been scraped bare by wildfire three times; and a young Indigenous video artist whose work draws from these mountains and her ancestors.

Also, a Tibetan Buddhist lama and neighbor who teaches me about interbeing; a woman who survived violent abuse, became a paratrooper, and hopes to remake her life by helping other women remake theirs; a tribal council member protecting ceremonial land; a retired fire chief who confronts wildfire with a garden hose and luck; and a nearby neighbor who, while fighting in the swamps of Vietnam, picked up a wounded egret and carried it into the future.

Casting a net beyond this region, I've interviewed artists, teachers, botanists and other biologists, writers, ranchers, photographers, conservationists, religious leaders, a visiting expert on "the loneliness of trees," and many others.

The term *deep noticing* does not represent a new concept. It has been practiced and taught in various forms, by different names, for thousands of years across cultures.

Now a new surge of research suggests the existence of a deeper, more mysterious network of life on Earth. Humans possess far more senses to perceive this network than Western society has generally acknowledged. Drawing on ancient wisdom and modern science, we're rediscovering that the forest is enchanted — and that we can become *bioenchanted*.

A friend, Janet Keating, recently sent me a note addressing these ideas.

Janet has devoted her adult life to fighting mountaintop removal in West Virginia. She wrote, "Deep noticing — isn't it about seeing with the heart? Connecting on an emotional level with nature, being present, accepting what we sense rather than only what we first see? It's a reciprocal relationship, a dialogue. We seldom find what we're *not* looking for. God is hiding in plain sight."

I hope this book offers new tools and inspiration to build a deeper relationship with the natural world and to find hope and healing, no matter where you live. In the meantime, I keep walking, hoping to notice more, to know enough to know when to come home, and to know where home is, even as it changes.

Here is good but fragile news. Counterintuitively, even as we mourn the destruction of so many species, lands, and waters, the natural world offers us an antidote to our growing despair. We can find relief right there in our peripheral vision, in a deep and wavering current followed by native trout, in a line of swifts living most of their lives in flight, in the tracks of a mountain lion along a worn path.

A path of wonderment.

I. WHAT ARE WE MISSING?

I wonder how many people I've looked at all my life and never seen.

— JOHN STEINBECK

First Place

LYING IN A CRIB, waking in silence, you notice the moth on the amber wall, or the face approaching. You lie still and watch the dust motes fall in light filtered through stained glass or a yellowed pane marked by long-ago raindrops. You sense the light behind the light.

And then a doorway opens to the world beyond, the streets and fields and woods and alleys and schools. To water and sky.

This imprinted First Place travels with you for the rest of your life. It serves as a kind of ark, carrying animals, plants, the curve of hills, and the rise of dawn above the trees.

It might have included a backyard garden, a circle of trees at the end of a cul-de-sac, an urban park, a path deep into the woods. There, you might have inspected a dung beetle as it rolled manure across a field, or the sunflowers reaching for the sky, or the old toad on the sidewalk watching you with red-rimmed eyes, or a crawdad drifting in the current of a creek, or ants moving the tiniest of leaves like foot soldiers on the way to the front.

Being new to the world and closer to the ground, children notice such details and create stories about them, or they paint or draw them.

Some of us, the lucky ones, hold on to that magic. Maybe you had a special spot in nature when you were a kid. Where every leaf felt new, and even now, you can close your eyes and be there again. Others spend their lives searching for a place in nature. Sometimes, they find it.

My earliest memories are of my grandmother's Victorian house, built by my great-grandfather in 1889 in Independence, Missouri. My grandmother, Edna Mifflin Streeter, was born in 1884, just nineteen years after the assassination of Lincoln. Her husband was born in 1863, when Lincoln still lived. In her forties, she gave birth to my mother in 1924. Nine years later, he died. Great elms grew in the front yard and eventually towered over it.

My grandmother was a kind and quiet widow who took in boarders during the Depression and World War II. Later, she took care of me when my parents were at work. I don't recall her ever saying anything unkind to or about anyone.

Her house and neighborhood were filled with real or apocryphal ghosts.

One day, my father was gardening in my grandmother's yard. Sitting under a pear tree, I watched him dig into a small earthen mound. Shards and broken arrowheads practically spilled out. The mounds were probably middens left by the Osage people as they traveled through the area long ago. Later, I imagined them traveling as spirits.

A forbidden ravine at the end of the road held its own stories. As a four-year-old, I was drawn to its edge. Far beyond my allowed boundary, I peered into the green darkness below. My collie, Banner, was by my side. To avoid sliding down into the ravine, I reached out for his mane of fur.

At that hour, the ravine was dark and silent except for a bird or insect call — whether complaining or celebrating, I couldn't tell. I absorbed the details of that place but did not want to know them all — perhaps because I noticed the smell of garbage and saw the rusting automobiles hidden among the green-and-black foliage. Or maybe because I could not see the bottom. Down there, trees and bushes fought for sunlight, their branches tangled in gnarly ropes. In my imagination, these were Tarzan swings. Banner wanted us to leave this place. I could tell by the sound he made: a huff.

Perhaps the vines were wild grape, rooted woody lianas that climbed the trees to reach sunlight in the canopy. The leaves are said to taste like grapes. Or maybe they were moonseed, easily mistaken for wild grape. Despite the threatening darkness, I wanted to climb down there and follow a path into other worlds. Then came a rumble and groan of a freight train riding the tracks on the far side of the ravine. Lights from the train pulsed through the vines, and I felt the cool air pushed up from the darkness below.

Through the years, we live in newer places, and if we're lucky, we bond with them. All these places become the geography of our lives. But there is only one First Place.

WHEN I WAS SIX, my family moved to the edge of suburban Raytown, Missouri, eight miles south of Independence. With Banner, I explored the fields and woods behind our new house. As we ran through the high corn, he raced ahead or circled me.

In the early summer, the leaves and ears of corn were green and sticky and prickly. By fall, the stalks were brown and brittle. No accident that so many of Stephen King's chapters take place in cornfields. Anything can happen. In a cornfield, at least for a child, the world is taller, the way forward less clear, the sounds muffled.

In the clearing of my woods, a rivulet of water flowed through the marsh. When the water reached the trees, it grew into a creek. I would venture into the trees to sit on the creek bank and watch the minnows, frogs, and water spiders do their dipping and sliding dance.

Thinking of that place is like remembering the back of my mother's gentle hand.

In the mid-twentieth century, there began a generational disconnect from nature. Even then, only a handful of kids in my neighborhood spent much time in the woods. In those years, I had no idea how lucky I was or knew that so many kids my age had no chance to climb a tree or sit in the clumped grass of a clearing in the woods. This is not to say that cities contain no natural habitats, but they compete with traffic noise, exhaust, and thousands of other distractions.

I missed out on the advantages some urban kids enjoyed. There was no easy bus ride to the American Museum of Natural History or Brooklyn's

Prospect Park or, for that matter, Kansas City's Swope Park. Such places could also have ignited an intense interest in nature. But for me, nearby nature offered a doorway to the world of the senses.

One day, I stood alone in the clearing. At the center was a marsh with islands of clumped grass. Long blades pushed up and out from the center of each clump. I judged each clump to be about the size of a sitting African lion, its mane rising two feet in the air. Not knowing its real name, I called it lion grass. With my feet in an inch of water, I turned and faced the cold spring wind. I stood still for a moment, leaning slightly forward. And then I threw myself backward into one of those clumps of grass. I felt like I was falling upward.

Settling into the blades, I lay there for a long time listening to the oak limbs sway. Watching a blue sky streaked with white and gray, I felt lifted on the wind, up and out of the lion grass.

Growing up, I often experienced what the ecophilosopher David Abram calls the "spell of the sensuous" or what Henry James referred to as the "palpable present-intimate."

When I climbed a tree, I could hear the wind fluting through the thin, long crevices in the bark. Looking down, I would see patient Banner looking up at me.

With legs and arms around the trunk and feet stepping on branches, I shimmied upward. In the wind, leaves whipped against my eyes and cheeks and ears and neck. I would study the bark and its colors and investigate the furrows populated by nearly transparent spiders. As I climbed higher, the tree groaned and swayed. I would sit up there for an hour or more. I would put my ear to the bark and try to hear those little civilizations.

These were the moments when the earth stopped.

In the clearing of the woods, a rivulet of water flowed through the marsh of lion grass. When it reached the trees, it became a creek. I watched the minnows, frogs, and water spiders do their dipping and sliding dance. Long ago, mountain lions or cougars disappeared in that part of the country, but I dreamed of seeing one. Or even a bobcat. Oaks and cottonwoods lined the creek and the edge of the woods.

I remember every scent, every brush of the air and the feel of the earth beneath my feet or between my fingers. Adults tend to focus more on the big picture. Kids are closer to the source of things. Even today, I can feel that lion's mane holding me, my fingers gripping its fur, and in those brief moments of remembering, I am again part of everything.

WHEN I WAS ELEVEN, our family moved again, across the border into Kansas, to a house near a lake, and during the move, Banner was hit by a car and killed.

It was a gray year. Sometimes in winter, I would get off the school bus, head down to the shore, and walk across the ice toward my house. Later, in the spring, I would follow the shoreline until the water narrowed into a stream and then a creek. Carp swam in the shallows and rolled in the sun in flashes of gold. Shadows moved beneath the surface. There were snakes in the swamp and in a pasture that ran along the creek. Beneath rocks and logs, I would find smaller brown snakes with gold rings around their necks.

One afternoon, a girl my age got on the bus wearing a coat with a faux-fur collar. She was full of life. I was thirteen, and my first thought was, *She's like a movie star.* Over time, I got to know the girl. I told her about my dog Banner and about the woods and the stream and the ring-necked snakes. She became a friend. Wanting to show her the place I loved, I asked her to go there with me one day.

She was not someone who often stepped off the sidewalk, at least not this far.

We walked through the hills and stopped at a barbed-wire fence. The shadows of clouds slowly moved over the trees and across the pasture. We could see horses down there, bounding, playing, their backs curving. The wind was blowing; the grass was bending. I looked for one of those clumps of lion grass and did not see any.

We stood there for a long time, both of us looking across the field to the woods, and she finally said, "I think I get it."

"Get what?" I asked.

"You know," she said.

And then we walked all the way home under the bright sun that had opened on that pasture.

CHAPTER 2

⯬ To Notice and to Be Noticed

D ECADES PASSED, but the spell of the sensuous did not.

Thinking about those woods now, I realize how often I was there alone, but I never felt lonely. Banner kept me company, of course. But any isolation I might have felt was balanced by the trees, the grass, the clouds, the birds, and all the wildlife surrounding me.

In one way, intimacy can be defined as the ability to notice and to be noticed. Or, to love and be loved. Most children are born into this state of attentive consciousness: to intuitively notice, to study not only the surface but also beneath and beyond that surface, and to imagine. Science requires such attentive consciousness. So does art.

This does not mean that growing up surrounded by nature can reliably predict a later sensitivity to the natural world. Or that the natural world cannot be experienced in a city. Or, for that matter, that a child or an adult, even in the wildest environment, will be open to nature at all. But circumstances can create pathways for deep noticing, especially when parents encourage a connection and for children from families in pain; these children may find refuge in a forest or a patch of vegetation in an alley or at the end of a cul-de-sac.

There, the child finds relief from loneliness, shelter in the company of other living things, and at least the sense that they are being seen.

The word *noticing* describes that instant of perception. The word *attention* suggests focus and being ready for input. Jan van Boeckel, a Dutch art professor and ecophilosopher who is also a friend, asks, "What is the difference between noticing and paying attention?" Then he answers his own question. "I believe noticing comes first. Noticing is a readiness to allow the new to reach our senses. Then we pay attention." He quotes Thoreau: "If you would make acquaintance with the ferns, you must forget your botany. You must be aware that nothing is what you have taken it to be."[4] Deep noticing is about the pursuit of deeper meaning in the beautiful and the mundane; it is a form of prayer, even if that prayer is only bearing witness to a tragedy. Or to a rebirth.

Deep noticing can improve our physical, mental, and spiritual health and attaches us more deeply to place and to our neighbors of all species. It can illuminate and deepen known science and lead us to ask better scientific questions. It can enrich and inform our art and music and architecture. It can build compassion and our sense of reciprocity with other kinds of life.

In 2016, researchers at St. Catherine University studied people as they immersed themselves in nature and reflected on their experience, using photography and journaling. The researchers' conclusion: "Regardless of the type of nature activity, the location, or the time of day, nature immersion offered a sense of connection for every participant, which in turn promoted spiritual well-being. Spending time in nature offers a cost-free and accessible way to foster spiritual well-being."[5]

D. H. Lawrence goes further: "Prayer, or thought or studying the stars, or watching the flight of birds, or studying the entrails of the sacrifice, it is all the same process, ultimately: of divination…Every real discovery made, every serious and significant decision ever reached, was reached and made by divination. The soul stirs, and makes an act of pure attention, and that is a discovery."[6] Such attention is an affirmation of life. And, I would argue, hope.

Pablo Picasso famously said, "It took me four years to paint like Raphael, but a lifetime to paint like a child." Which is why returning to the First Places of our childhood, in memory or in person — whether in suburban woods, a farm, or a city park — can be so revealing. The remembered experiences can reignite our abilities to notice and help us understand our chosen places now.

AFTER COLLEGE, I MOVED from the Midwest to San Diego. I lived in a cottage overlooking a beach. The ocean view enthralled me. After a while, I had to remind myself to look — to notice the ocean. Habit dims sight and sound. "Only the blindness of habit convinces us that we continue to live in the same place, that we see the same landscape," wrote the Irish poet and philosopher John O'Donohue.

Most days, I would walk to the end of a long pier that reached beyond the breakers. Now and then, I would watch dolphins or orcas arc over and through the waves. I loved the ocean, but for the first couple of years, the city still felt alien. I missed box turtles, cornstalks, and even the pea-green clouds of tornado season.

One day, I drove east on the freeway beyond the city's edge and saw a dirt road to my left, winding higher into hills of chaparral.

I exited the freeway and found the entrance to that dirt and rock road. I drove north for miles, passing the granite flank of El Cajon Mountain (more commonly known as El Capitan or El Cap). I learned later that the mountain is sacred to Indigenous people who have lived there for ten thousand years or more. Once, the valley held Kumeyaay farming villages.[7] After the villagers were forced out in 1932, reservoir water from the Cuyamaca Mountains was sent crashing down through thirty-five miles of wooden flume, crossing 315 trestles and eight hand-dug tunnels, filling the valley, and drowning the abandoned villages.

As I looked down at the lake, a hot Santa Ana wind sliced up from the eastern desert, then down from the mountains, shaking the car. I kept driving but had no clue where I was going. I didn't know about the lost villages or the old mining town of Julian yet. I steered to the top of a hill overlooking the lake. From there, I could see higher mountains to the north, east, and south.

On this open dome, tall brown grass spread out in a kind of rolling prairie, flattened by the rising hot wind. Stopping the car, I got out and walked among clumps of grass.

I came to one clump that was about the size of a lion, and I turned in the wind. I vaguely remembered that moment in the meadow when I was seven years old. Instinctively, I put out my arms, faced the wind, and fell back into the embrace of that grass.

FOR YEARS, I FELT DRAWN to wherever that wind came from. Even after San Diego began to feel more like home, I felt that pull. And a growing hunger.

Generations of newcomers have sought that something in these mountains. Some of them may have been the first people in North America, according to a controversial study conducted by a team of paleontologists from the San Diego Natural History Museum and published in the journal *Nature* in 2017. The claim was based on markings on mastodon bones discovered near San Diego. If the findings are accurate, the study "would significantly alter our understanding of how humans spread around the planet," the *New York Times* reported.[8]

In addition to being inhabited by Spanish colonists, tech workers from Ohio, and numerous Kansans, San Diego County is home to the Ipai-Tipai Kumeyaay, who traveled down the coast by sea, island to island, to the mainland. The Payómkawichum, the Kuupiaxchem, the Cahuilla, and, down the eastern slope of the mountains, the Kamia, a Yuman-speaking people, were also early inhabitants.

These and other groups are all part of the soil, water, and trees here.

The website of the Viejas Band of the Kumeyaay features a description of the clay, pigments, and minerals that play a significant role in traditional Kumeyaay culture and cosmology: "Clays and pigments are not seen as just physical elements — they are part of a deeper understanding of the earth and the land. This is reflected in the fact that the Kumeyaay word for land/earth and the human body is the same: Mat/'Emat."[9]

Whoever lived beneath the Volcan shadow first, each generation of newcomers reinterprets the meaning of these mountains and the valleys that link them.

Locals call Volcan Mountain a mountain, but at thirteen miles long, it's more of a small mountain range. From a distance and an angle, it does appear as one long peak. Volcan is part of an archipelago of isolated mountains that begins in the southern Sierras and extends southward through Riverside and San Diego Counties into the Baja California peninsula of Mexico.[ii]

Some of these peaks and ranges are sky islands, defined by their physical and biological isolation from the land around them.

South of here, in Baja, the white granite Picacho del Diablo (Peak of the Devil) rises 10,154 feet, towering over much of the Sierra de San Pedro Mártir. In *Fremontia,* a journal of the California Native Plant Society, botanist Thomas A. Oberbauer calls Picacho del Diablo "ethereal...primeval..." Some of its species are relics of the Pleistocene (also known as the last Ice Age). Wildlife and plants migrated to Diablo in wet times before the climate became drier, and, as Oberbauer writes, "these mountains must have been cut off from the north."[10] A visit to a true sky island, Oberbauer writes, "provides a window to the past."

Plants and animals on the high slopes of Volcan may not be as exotic as those found on Diablo, but Volcan "provides a stronghold" for rare plant and animal species, according to the Volcan Mountain Foundation.[11] The first species of rainbow trout — progenitor of many evolved descendants that have traveled around the world, often planted by British sailors who carried them in whiskey barrels — may still exist in this county. Incomplete biological studies have recorded 45 species of mammals on Volcan, including flying squirrels, mountain lions, occasional black bears; and over 150 birds, 36 reptiles, 12 amphibians, and an unknown number of insects and other invertebrates. The montane forest offers refuge but also isolation. Mountain lions, cut off from their kin to the north by highways and housing developments, are starting to show signs of inbreeding, including the occasional appearance of oddly crooked tails.

By a broader definition, sky islands exist everywhere — in the dust bowl windbreaks planted to protect farmland on the plains; at the ends

---

ii   East of here, a string of better-known sky islands reaches south through the Sonoran Desert regions of Arizona and northern Mexico.

of cul-de-sacs and in ravines of suburbia; even in the corners and crevices of inner cities — anywhere life finds refuge, especially as climate change advances. Some of these islands will disappear by the time you finish reading this sentence. Large or small, a sky island can become a graveyard of dead ancestors or a fertile mother bursting with new life and hope.

IN ADDITION TO WARNING his readers to guard against losing sight of beauty because of familiarity, John O'Donohue wrote this: "Gathered high in silence and stillness, the mountain is loaded with memory that no mind or word can reach."

In stillness, deep noticing builds from a sense of place — or "placefulness," as artist Jenny Odell puts it in *How to Do Nothing: Resisting the Attention Economy.*[12] She defines *placefulness* as a "sensitivity and responsibility to the historical (what happened here) and the ecological (who and what lives, or lived, here)." You can feel placefulness in Soho or Riverside Park or LA's Venice Beach, or in the cluster of trees lining a suburban sidewalk. For a child, those trees can become a universe. For all of us, natural places can become *attention sanctuaries* — places we intentionally find or create for conscious withdrawal from the influence of technology and for focusing on the neglected parts of our lives.

Others are drawn to the practice or sensation of placefulness.

Jason M. Brown, a lecturer in religion and ecological humanities at Simon Fraser University in British Columbia and author of *Dwelling in the Wilderness,* writes about what he calls "contemplative ecology." The concept of placefulness, he says, seems "to sum up in a simple word how monks in the Benedictine/Cistercian monastic tradition related to the landscapes of their respective monastic communities. They were present not just to abstract theological notions, or the love of God, but to the intricacies of their surrounding environments."[13]

In this spirit, Sarah Elisa Kelly, a lecturer at Schumacher College in the United Kingdom, speaks of the "many pathways we might take towards feeling ourselves more place-full."[14][iii]

---

iii   Some who use this term spell it *placeful*; others use *place-full* or *placefull*.

To stretch, to deepen our relationship to place, Kelly offers this suggestion: "A useful exercise I've been experimenting with lately comes from the developing field of nervous system and trauma-based research…It involves learning how to 'bridge' our attentions." She suggests becoming aware of "the points of contact, where your body touches the floor beneath you, or the chair you sit on. Notice what your hands are touching; perhaps they rest on your legs or a wooden desk. Feel the softness, scratchiness, hardness, and warmth…then include the vision, sounds, and smells if they are also present." Further, Kelly advises, be "attentive to detail such as the folds of fabric, the translucence of a leaf."

The sensation of bridging all these layers can be "a kind of shimmering fullness, an experiential embodiment of the aliveness of 'place.'"

The Green Earth Foundation, a nonprofit organization working to change public attitudes about ecological balance, offers a list of starting points for developing a deeper understanding of where you live or are thinking of living:

What are the weather patterns and natural hazards in the bioregion?

What is the soil like? What kind of rocks and minerals are found there?

Can you name native plants or trees in the region and their seasonal availability?

Can you identify birds or any other animals resident in or migrating through the region?

Do places in the area have a special attraction or power for you?

A similar list of questions is suggested by pioneering ecopsychologist Patricia Hasbach, a retired professor at Lewis & Clark College in Portland, Oregon, and the author of *Prescribing Nature: A Clinician's Guide to Ecotherapy*. For "place bonding," she recommends going somewhere in wild nature "once a week, or more, several weeks, at various times of the day and in different weather conditions…Let yourself settle in…relax…breathe…and open your awareness to the surroundings." Then ask yourself: "How do you feel in this place? What draws your attention? What do you see, hear, smell, and feel? What are some of the relationships you observe in this place? What is the nature of your relationship with this place? How has that changed over time? How do you experience, if at all, healing in this place?"[15]

Understanding how Indigenous people connect to the place, now and in the past, will add an essential and deeper level of meaning. Carl Jung, the Swiss psychiatrist and psychotherapist, argued that people would not be so "divided against themselves" if they had lived in a time and place where human beings were "still linked by myth with the world of the ancestors, and thus with nature truly experienced."[16]

But what if we do live in such a period, and have yet to notice? Consider what we still have: a planet that turns; our instinct to love life, people, and places; and the great conversation that circles us in the trees, beneath our feet, and through time.

Can we hear those voices?

Nearly 9 million animal species live on our planet. Of these, only 1.2 million have been identified and cataloged. There are about 435,000 unique land plant species; 611,000 species of fungi; some 3.2 trillion species of bacteria. And among the protists — organisms that don't quite fit into the three eukaryotic kingdoms and that range from single-celled organisms to multicellular seaweed — there are nearly 200,000 species of algae and other protists.[17]

These tallies don't include rocks, mountains, and rivers, which, if you follow certain spiritual or scientific beliefs, are also, in a sense, alive. The strongest case for a broader definition of life comes from members of Indigenous tribes, but many traditions and even scientists explore this idea as well.

Usually, we ignore the great conversation, but it's still present even after a snowstorm in March — maybe especially then. Listen.

SOME PLACES VIBRATE with an extra sense of life. Your feet are drawn there, then your heart.

Kathy and I often headed to Julian and the surrounding wild areas when we needed to escape the city. Memories were made there.

One time, Kathy, the boys, and I tromped around in a foot of snow after a March snowstorm. This was the first snow that either boy had ever seen. Jason and Matthew seemed born to it. They threw snowballs and rolled big boulders of snow to make a snowman and a snow cave. As light faded, Kathy and I watched the tops of their knit caps from a back-room window and listened to slabs of wet snow and ice drop from the high branches.

The next morning, we put on our boots and snow clothes and drove to a trailhead. From the edge of Laguna Mountain looking north, we could see the peaks of the San Jacinto Mountains and the top of the Volcan range, white with snow. At the edge of the mountain, we looked down at the saucer — or maybe a frying pan — of the Anza-Borrego Desert, almost six thousand feet below. The two environments, so alien to a Kansan, looked like planets colliding.

On the way back to the cabin, I led us off the trail into deeper snow. Jason and Mathew insisted we were lost. "Daaaad," they said.

"I know where we are," I answered. But I listened for cars on the road below, just to make sure.

A few years later, I needed some solitude to meet a deadline. Artist and sculptor James Hubbell and his wife, Anne, were friends of ours. For forty years, they had worked to create a compound of art studios and small buildings that looked like they had been sculpted from stone. Artists from around the region would come there to do their work and to be inspired by Anne and James. The compound was in the mountains, with Volcan Mountain to the northwest. It would be empty for two weeks, and they needed a house sitter. Aware that I was facing a deadline, they invited me to stay at what they called the Boys' House and work there. Just remember to feed the dog and cats, they asked.

I drove up alone. From the doorway of the Hubbells' Boys' House, I could see the dark Cuyamaca range and the valley below, where horses ran through fields. The compound was almost in the shadow of Volcan Mountain, a mile or so to the north.

I planned to work here for a week. Kathy, Jason, and Matthew would visit midweek.

On the first day, I wrote until late and then went to bed. I dreamed noisy urban dreams. In the slate-gray predawn, I opened my eyes and saw the face of a coyote a few feet from mine. It stood on the other side of an open stained glass window.

The coyote stared at me. I blinked. It was gone.

During the next few days of solitude and work, the moving clouds and lifting wind brought voices — of Kathy and my sons, whom I missed, and my father and mother, both now gone.

On the fourth day, I heard Kathy's old Volvo winding up the road.

Maybe it was the heat or the strangeness of the surroundings, but initially we all felt a little reticent and awkward. Perhaps it was the solitude, but familiar phrases and patterns seemed odd, somehow.

At the end of the short visit, Kathy took me aside and said that our younger son would like to stay with me for my remaining three days. He was bored at home and needed a break from his brother (and his brother needed a break from him).

"Of course," I said, "as long as he understands that I need to work, and he'll have to entertain himself." At eleven, Matthew was between childhood and true adolescence. My wife and older son drove off, and Matthew and I went through the house looking for books for him to read. He finally picked out *The Fellowship of the Ring,* by J. R. R. Tolkien.

I flipped through the Tolkien book and read a few lines: "That night they heard no noises. But either in his dreams or out of them, he could not tell which, Frodo heard a sweet singing running in his mind: a song that seemed to come like a pale light behind a grey rain-curtain, and growing stronger to turn the veil all to glass and silver, until at last it was rolled back, and a far green country opened before him under a swift sunrise."

No television, no ringing doorbell, no homework. Matthew sat on an old couch behind me. Respecting my need for quiet, he began to read. Three hours later, I realized he had not said a word or made a sound. I turned around.

He was asleep, holding the Tolkien novel like a stuffed bear.

That evening, we walked up the hill and swam together in a round, tiled pool under a quarter moon, and later, we listened to the wind come up and the coyotes jabber in fits and starts. For the next three days, we talked only occasionally, usually in the pool or during dinner. He is a voluble boy, with many comments to make about his world, so I was surprised when silence came so easily to him. The absence of electronics, except for my laptop computer, certainly helped. So did the wildness of the land around us.

I asked him to take charge of feeding the dog and the cats. He gave names to the cats — Crusty, Tootles, Frisky, and Shady — who followed him around the property, scrambling up the oaks to show off for him. In the evenings, Matthew and I took long walks through the woods and down to the pastures and orchards. With his Kodak camera, he sneaked up on the deer that wandered through at dusk.

We moved to a rhythm unknown to us before. I got to know Matthew better during our days in the Hobbit — or rather Hubbell — house, overlooking the far green country, and perhaps he came to know me better, not because we talked but mainly because we didn't. As a parent, you capture such quiet moments when you can, in the loudness of time.

THAT WEEK'S STAY was part of a series of events that ultimately led to our decision to make these mountains our chosen place.

One day, Kathy decided it was time. Good timing, too. This was a year before the pandemic. We moved thirty miles east of San Diego to the outskirts of Julian, a former gold mining town settled in the 1870s by formerly enslaved people and former Confederate soldiers. Today, depending on the month, Julian holds a population of 1,491, give or take.

During our first year, we rented a house in town while searching for a house to buy. Then, in 2019, we found the house we live in now.

Our closest neighbors include turkeys, bobcats, ringtails, opossums, rattlesnakes, ravens, crows, red-tailed hawks, golden and bald eagles, silver-haired and big-eared bats, and a retired fire chief. Our trail cameras capture images of night visitors in the act of living: the purposeful coyote and the pregnant opossum. On our acre, native and invasive plants fight for territory. We side with the locals.

And Kathy and I try to walk four or five miles a day. Emphasis on *try*.

She moves with serenity. I poke at mud and rocks and logs and tracks with my hiking stick, like a turkey. One morning, at the top gate, I asked her which way she preferred to go. "That way," she said, pointing east. I was surprised. I knew she was tired of walking the same old route, the one I typically followed. On that morning, she picked my same old route.

"Why that way?" I asked.

"Because I know you'll want to see if anything has changed." She smiled. Kathy knows I don't like change, and because I dislike it, I notice it.

Pausing, we watch a doe, its nose in the grass at the end of our acre. She looks pregnant. The doe lifts her head and watches the morning sunlight emerge through the branches. Now our eyes track a turkey gang (one of the correct terms for a flock of wild turkeys, I swear) as they peck their way north. In this part of the country, wild turkeys are not native, but they're still welcome.

Every few years, a bear will wander down here from the San Bernardino Mountains due west of Los Angeles.

Our home is surrounded by one of the largest concentrations of humans in the world. San Diego and the Pacific Ocean bracket us to the west. Los Angeles, the nation's second-largest city, is to our north. Tijuana, Mexico's second-largest city, is seventy miles to the south. And yet, nature survives, and here and there it thrives.

In 2024, the PBS series *Nature* named urban San Diego America's Wildest City, a name that doesn't make sense until you realize how diverse the region's climate and geology are.[18] San Diego County ranks as the most biodiverse county in the continental United States. It's home to "200 imperiled plants and animals — more than in any other county in the nation," according to the Nature Conservancy.[19]

This county also claims more organic farms than any county in the country and is home to more Indigenous tribes and reservations — many of them small — than any other US county.[20] And down the back slope of these mountains from us is the Anza-Borrego Desert, one of the most remote places in the state, much of it accessible only by private roads or on foot.

All in all, it's a good place to notice.

Week after week, our mental map of the area grows. Volcan Mountain fills our front windows, in the winter held to the earth by a cloud of snow. Fog, like white lace, moves in and across the valley below us, obscuring everything except the peak, and then that disappears. Great clouds boil up from the Anza-Borrego Desert, loom above Julian, their bellies full of darkness, then move on. Shadows slide across the valley, painting the mountain red.

Volcan seems alive. Like us, the mountain changes from moment to moment. It has stories to tell. Scenery is summary, the CliffsNotes.

CHAPTER 3

In the Gaze of a Mountain Lion

JUST AFTER DAWN, Kathy was upstairs at her computer. Our two cats were stretched out on her desk, holding down her papers. Suddenly, Piccolo Pete growled. A second later, Mister Hightop, our alpha cat, leaped from the desk and fled to the basement.

Looking out the window, Kathy saw a large mountain lion walking up through our west side yard toward our house, holding something brown — probably a ground squirrel — in its jaws. The lion moved calmly and majestically past the deck and up the stone walkway to our gate.

Pushing past the entrance, it turned left and walked along the road at eye level with Kathy's office window and then disappeared.

At this time, Piccolo Pete also decided a strategic retreat was appropriate. He shot from the desk and joined his brother in the bunker. I slept through everything.

In and around Julian, it's rare for anyone, even people who have lived here for decades, to see a mountain lion in the flesh other than in trail-camera photos.

You seldom see what you're not looking for.

Our trail cams are usually on the east side because that's where the deer trail is. But when Kathy saw the mountain lion from her office window, I moved all the cameras over there. The next morning, I saw the lion's image. Large and muscular and located about forty feet from the side deck, it seemed to flow across the grass and stone. A friend looked at the photo later and described it this way: "That cat looks like a *railroad*."

On a Saturday that fall, a ranger associated with the Mountain Lion Foundation, a nonprofit that monitors the cats' welfare, travel patterns, and human interactions, spoke at the local library. Big turnout. Most people in the audience that night lived in Julian or the surrounding territory. Later, many of them could only remember one thing the ranger said.

"How many of you have seen a mountain lion?" the ranger asked.

Five people raised their hands. The ranger smiled. "For the rest of you, I can just about guarantee that a mountain lion has watched you. You just never knew it."

Doug Tallamy, a botanist and writer at the University of Delaware, confirms this statement. He tells me that mountain lions are known to watch people from the cats' secret places and then emerge to investigate the humans' tracks, their refuse, and their scents.

I wonder, What do wild animals know about us? Sense about us? Do they spread rumors?

More than any other critter here, the mountain lion haunts the local imagination. I have heard people say they moved here because of the mountain lions. Shortly after Kathy and I settled here, I was standing in the parking lot of the Volcan Mountain Foundation's education center talking with some neighbors. We were trading mountain lion stories — sighting from afar, tails disappearing into the brush, the thump on the porch. We grew more animated as we talked. For a few minutes, there was no room for rumination about the news or the state of politics. Or, for that matter, about ourselves.

Janice Bina-Smith, our Julian neighbor, returned home from a twilight walk in the neighborhood. "I got in the door, and several minutes later, I hear a man's voice outside yell 'Hey!' really loud," she recalls. "My house has a steep pitch on one side and a lesser pitch on the other. I heard something jump on the back of the roof, and I could tell it was larger than any other animal I've ever heard up there before."

She listened intently to the sound of four paws climbing up the steep pitch to the top. "The shingles on the roof were crackling, and I could hear claws taking hold." No other local animal would make those sounds. "My cat and I were both looking up at the ceiling with our eyes wide open," Janice recalls. She felt a sense of awe mixed with a tinge of fear, or maybe amazement. She continues: "When it reached the top of the roofline, it slowly walked down the other side to where the porch and front door are located. Right over my head. The roof ends probably ten feet above the ground, and the footsteps stopped right at the edge."

She considered opening the door to get a look at it. "Then I thought, 'Are you *crazy?*' I kept thinking that maybe I should call somebody," she recalls. "And then I realized that the only thing to do was to let it find its way down by itself." She also sensed that the lion was afraid and was running from someone and had jumped onto the roof to escape. "Then the lion began to walk around on the roof as if it were trying to figure out its next move."

Dark was coming on. The footsteps stopped.

"I thought that it must've jumped off the roof," she remembers. "Then, as I stood by the windows on the side of the house, I saw a dark shadow walk past me." And then it was gone.

One night not long afterward, Janice had just turned out the lights and gone to bed. Suddenly, her house cat hurtled into the bedroom and hid in the closet.

Janice tiptoed downstairs. Through the living room window, she saw the mountain lion, perhaps the one that had walked on her roof. It was drinking from her birdbath.

"The sound of that big tongue licking up all that water was so loud!" she says. "The lion turned and sat in front of me on the corner of the porch, looking at the mountains. He sat there like the Lion King. Then he decided to take a bath. Just like any cat, he licked himself all over...It was like witnessing a miracle."

What is it about our fascination with—or fixation on—lions and tigers and bears? Why the awe? Is it the strange excitement of being lower on the food chain, of being — almost hoping to be — stalked by a ghost in the forest? One reason is that a sense of humility is a survival tactic.

"For me, it helps to remember that there are many storylines, from top to bottom, other than my own," a friend said the other day. "Keeping things in perspective that way is strangely comforting." Humility. That odd sensation. Plus, we like the excitement.

In his book *The Practice of the Wild,* poet Gary Snyder comes as close as anyone to describing the mystery of predator and prey. "Life in the wild is not just eating berries in the sunlight," he writes. "I like to imagine a 'depth ecology' that would go to the dark side of nature — the ball of crunched bones in a scat, the feathers in the snow, the tales of insatiable appetite…Life is not just a diurnal property of large, interesting verte-brates; it is also nocturnal, anaerobic, cannibalistic, microscopic, diges-tive, fermentative: cooking away in the warm dark."[21]

Sometimes referred to as the ghosts of the forest, mountain lions hint at a world beyond ours. We feel the "presence of something vast that tran-scends our current understanding of the world," says Dacher Keltner, a University of California, Berkeley, professor of psychology and one of the leading experts in the study of awe.[22] Many of us are drawn to or even love large predators in part because we fear them. Like beauty, fear is often an ingredient in the experience of awe. We are, in those moments, out of our comfort zone, no longer in control.

After all, a mountain lion can jump twenty-two feet vertically and forty-five feet horizontally — and if it's carrying a deer, sixteen feet horizontally. These Olympic climbers have been photographed on top of telephone poles. They can sprint fifty miles per hour. They can also swim well. And like your house cat, they often sleep for sixteen hours a day. Or more.

The house cat comparison has its limits.

DURING AN EXCEPTIONALLY DARK night, from the road behind the house, near where we had discovered a deer kill months before, we heard a moun-tain lion scream.

The sound, as often described, resembles the cry of someone being murdered.

I rushed around to find my iPhone to record it. Then I heard Kathy upstairs sliding the bedroom window open hard and yelling in her best mom/nurse voice: "You leave that alone! You put it down, *now!*" She had assumed that a small being, or a large one, was meeting its end. The lion's shriek stopped suddenly and never returned.

Once again, our cats freaked out; Piccolo was sure the screamer was under our bed.

According to family lore, my grandfather remembered that sound, which he had sometimes heard while working on the construction of railroad tracks in Colorado in the early part of the last century. He described how the screams crossed the plains in the night — and asserted that there was no sound quite so unsettling. Indeed, you can find multiple YouTube videos of mountain lions caught on trail cameras, standing rigid, jaws open, and emitting that unearthly sound.

Then there is the chirping. One evening, we heard loud chirping coming from the road, thirty feet from our open bedroom window. The sound was not identical to any bird we had heard. We learned later that mountain lions do chirp. Both females and males make that sound when they're calling for a potential mate, which will answer in kind, on their own dating app.

Mountain lions, bears, wolves, and jaguars remind us of who we are. Here's another reason for our attraction to them: We identify with the predator — the hawk, the wolf, the big cat. A prey animal also has traits to be admired. It lives in the moment, tuned for safety through vigilance, speed, and the ability to flee. Deer and horses seldom live alone. They keep each other within eyesight; they take turns as sentries while the others sleep.

Once, while walking on a long, straight stretch of road near our house, I stopped to watch two horses put their tails to good use. They were slapping flies away from each other's ears and hide. They took turns doing this. Our species could use such comity and community.

We try. But look in the mirror, writer Diane Ackerman advises: "Most predators have eyes set right on the front of their heads, so they can use binocular vision to sight and track their prey. Our eyes have separate mechanisms that gather light and pick out important or novel images. Prey, on the other hand, have eyes at the sides of their heads," because they need to notice if something is sneaking up on them. "Something like us," she adds.[23]

Anyway, who wants to be prey? Not us. We set the mousetrap, patrol the perimeter, shoot for sport. Better to be an eagle than a turkey. Following the ancient Greeks' advice, some of us do what we can to "tame the savageness of man and make gentle the life of this world." But we're only human, right?

A few months after we heard the mountain lion chirping. I was on another trip up Volcan Mountain with Judy Kemp to collect the memory card from our trail cam. I was driving this time, and when we rounded a curve, a mountain lion suddenly appeared in the road.

It ran right in front of the car and then around the right front bumper. It then leaped across the ditch and bounded up the hill into the chaparral and manzanita. Dark-butterscotch fur with black highlights and a black-tipped tail. Too fast to capture with our phones.

In a matter of seconds, a mythological, intellectualized, romanticized creature became flesh and blood.

II. THE NEUROSCIENCE OF NOTICING

In the woods, we return to reason and faith. There I feel that nothing can befall me in life — no disgrace, no calamity (leaving me my eyes), which nature cannot repair . . . I become a transparent eyeball; I am nothing; I see all.

— RALPH WALDO EMERSON

Secrets of the Super Bloom

A MOUNTAIN LION CONCENTRATES the mind, but a super bloom explodes it.

One April morning, after we had moved to the mountains, Kathy and I were joined by my boyhood pal Pete and his wife, Cheri, who had flown in for a visit. We drove through Julian on State Route 78, then descended into a rugged mountain pass to the Anza-Borrego Desert.

We passed a few trailer homes. A stream ran along the road. The mountainsides were coated with house-sized rocks, great gray marbles left behind by a receding ice age or by a thawing-and-freezing cycle that turns dirt into rock and squeezes the rock up through the earth.

Looking up, we saw the jagged fingers of lonesome ocotillo cacti. Then, below, where the valley widened out, a carpet of wildflowers stretched for miles.

We had arrived at a massive wildflower bloom called a *super bloom* (or "Flowergeddon," as a headline writer for the *San Diego Union-Tribune* once described it). This super bloom typically occurs between late winter and early spring. When the desert ignites, people arrive like honeybees.

Everywhere, you see desert trumpets, wild buckwheat, Texas filaree, and desert sunflowers; the low hills are coated with rainbows, and the desert washes are painted white by minerals sucked upward by sheets of water from the earlier monsoon.

At the lowest point in the desert, we stepped out of the car and walked among the wildflowers. Big green, yellow, and black larvae of the white-lined sphinx moth, sometimes called hornworms, plastered the ground and the creosote bushes — thousands of finger-length caterpillars were wriggling and climbing flower stalks and secreting acid when threatened.

In the distance, we saw clouds of ravenous Swainson's hawks with eight-foot wingspans diving into the endless buffet of larvae.

Year after year, these hawks arrive in the desert after a six-thousand-mile migratory trip that takes them first up the eastern seaboard of Mexico, then to the Great Plains, where Pete and I grew up. The rest of the birds climb the Pacific coastline to the Northwest states, where they nest. Their numbers continue to increase. Now, as many as twelve thousand hawks fly up the coast from Mexico, take a right, and head sixty miles inland to this spot, where they fatten themselves on caterpillars to build their energy for flight — and later, for sex with their life partners.

As the surrealist poet Paul Éluard put it, "There is another world, but it is in this one." Or, more accurately, otherworlds. Some exist in the most unlikely places.

Consider the desert's top crust, the dark, thin splotches here and there beneath the leaves and spines. Living organisms, including algae, lichens, mosses, fungi, and colonies of cyanobacteria, create that crust. Just add water, and the cyanobacteria produce gelatinous filaments that bind particles of soil, preventing erosion, reducing evaporation, and converting nitrogen into food for plants.[24] This crust on the desert floor is the superpower behind the super bloom.

Kathy walked along a dirt road through the flowers and caterpillars already flattened by three-wheelers. The caterpillars made their desperate, slow runs across the tracks. She picked up stragglers and carried them to temporary safety. In this cloud of swooping, crawling life, I felt bonded to the friends with us this day, to the hawks, to the caterpillars carried in the hawks' talons, to the sea of flowers reaching up, and to Kathy.

We head a half hour east to Fonts Point, where we walk to the edge of a canyon. As Kathy, Pete, and Cheri move farther along the ridge, I see a single banded rock lizard sitting on a knoll looking down and across a deep badlands canyon that stretches for miles.

Lizards' eyes detect a wider range of colors than do ours, extending to the ultraviolet range. They see a world that we cannot. I crawl toward the lizard to take a picture. He (assuming it's a male) is looking away from me. I wonder what he is thinking. Is he searching for flying insects or a mate? Is he as pleased as he seems — with the view, with himself, with life?

Then he's gone — in the air — making an impossible jump to another rock. I swear I can see joy in the lizard's leap.

I CAN IDENTIFY a lizard species at twenty feet. Or a mountain lion by its chirps. But local monkey flower or California fuchsia? Forget about it. I'm better at trees, but barely.

In short, I'm plant-blind. But wait. Researcher Kathryn M. Parsley of the Donald Danforth Plant Science Center in St. Louis, concerned about ableism — in this case, imbuing a term for a physical disability with a negative connotation — has suggested an alternative term: plant awareness disparity.[25]

By any name, PAD has been a concern to botanists and educators for centuries.

Ainara Achurra, a professor at the University of the Basque Country in Spain, defines PAD as a failure "to recognize the role of plants on earth and believing that plants are somehow inferior to animals."[26] Symptoms include not only a reduced ability to see plants outdoors but also a disinclination to understand the important roles plants play in animals' lives (beyond serving as food) and an unawareness of the different timescales on which animals and plants exist. Education is partly to blame. Research reveals that U.S. high school biology textbooks devote only 14 percent of their content to plants.[27] Not surprisingly, students prefer to learn about animals. Some critics call this an expression of animal chauvinism.

Cognitive and cultural biases override accurate perception. Humans are anthropocentric in layers; we tend to rank wild animals as inferior to ourselves, and plants inferior to animals and consequently less worthy of our attention. Some of us don't perceive plants as being alive, because they don't seem as lively as animals. They don't move around (though that assumption is being challenged). Also, plants are too, well, nice.

Humans are tuned to danger. We're not as motivated to identify plants, because they're not predators, at least not toward us. Our eyes seek mountain lions more than dandelions. Set aside the fact that a bush can also ruin your day. On one of our walks, Kathy and I notice poison oak plants along the side of the road. They have grown to the size of octopuses. Oily, green octopuses. The Venus flytrap and other carnivorous plants should also command our attention, though the genetically wily flytrap snaps its "jaws" shut faster than the human eye can detect.

Additionally, the human eye struggles to distinguish the morphology — forms and structures — of plants, particularly when different species of plants coexist in the same area and share similar colors or shapes. Some plants can "change their visual signal," according to Achurra; this ability helps them "avoid recognition and, thus, predation."[28] Other plants grow in places where they're hard to see or reach, such as slopes or off the beaten path, or "above or below the browse line" of plant-eating predators. Still other plants camouflage themselves by growing among less tasty lookalike plants. Or they favor the colors more common among spiny and poisonous plants: yellow, orange, red, brown, black, and white. Other plants hide in temporal refuges, growing or flowering when plant-eaters are off duty.

Trees seem to live in their own timescapes. Most people who do not see plants in their fullness also suffer from "time blindness," or a narrow interpretation of time's passage, writes philosopher Jonathon Keats. "A bristlecone [takes] hundreds of years to mature. Each organism's clock — its sense of time — is so different from ours that we can't even sense it."[29] Or we can. He describes how clockmaker Phil Abernethy, in his studio on the coast of Vancouver Island, is designing and building an "arboreal clock that won't show the minutes and hours of an ordinary human day. Instead, his clock will display time as experienced by some of the oldest trees on the planet."

Here are some ways to increase plant awareness: Grow a garden for food or pleasure; in education, rebalance the amount of time and attention given to plants and animals; explore what some consider plant sentience. Additionally, engage children by planting native plants in the yard or by greening school grounds; encourage nature journaling and botanical drawing. You can also learn about wildcrafting, the ancient practice of searching for and harvesting native and non-native plants for food, medicine, or crafts. And you can remove invasive plant species from natural areas and, where appropriate, work with others to plant trees and native plants.

And why not elevate plant identification to the competitive level of birding?

Each of these approaches to plant awareness can be adjusted and applied to the rest of nature, especially for people like me, whose eyes are drawn more to animals. Recognize the lizards and birds that depend on a particular plant, and interest in that plant goes up.

We need a larger, more inclusive term to describe how much we fail to see or sense. The correct term should apply to wildflowers, deer, mycelia, and mountain lions, all life on land and in water. Life awareness disparity (LAD)? Broad but bland. In any case, the main point is connection, not correctness.

For many, life is insular by choice; for others, the options are limited. In San Diego, the Ocean Discovery Institute, a nature education organization, conducted an informal study of children living in local urban areas. It found that approximately 90 percent of these children did not know how to swim, 95 percent had never been in a boat, and 34 percent had never been to the Pacific Ocean — less than twenty minutes away.

For others, deep noticing is a privilege, earned or unearned. But every child—and every adult — deserves a chance to flex their senses, to be part of the great conversation, to feel less alone on this planet.

Late that afternoon, Pete, Cheri, Kathy, and I headed back up the mountain. The low sun's light curved around naked red cliffs, blinding me at times. This road is perilous when you're headed west near sunset or when a thunderstorm or flash flood threatens. Danger gets our attention faster than beauty. But to be conscious, awake, we need both.

As we reentered Julian, I was thinking about that lizard leaping, wondering what worlds it sees from that high rock, what the earth looks like to a Swainson's hawk, and what concerns the caterpillar as it climbs toward the highest wildflower.

CHAPTER 5

Whispering Trees

NEAR THE SUMMIT of Volcan Mountain, Janice Bina-Smith sits with a group of us in a circle on crunching oak leaves. The red ants don't seem to be in a biting mood. Patches of sky above are clear. Gusts of wind move the trees, wagging their tips.

"The forest is the therapist," she says, "not me. I'm the guide. Your guide only opens the door." What we will experience is nothing like hiking, she adds. "When people are hiking, they're usually eager to reach a destination. They don't fully experience what they see. Forest bathing is a way to slow down, step out of our bodies. Stopping to notice everything."

Janice picks up a piece of bark, rolls it in her fingers, and smiles. "It does help to remember what you noticed as a child." A musician and nature educator, Janice (who told the tale of the mountain lion on her roof) is in her final stage of becoming a forest therapy guide. In a few weeks, she will be accredited by the Association of Nature and Forest Therapy Guides and Programs. She hopes to make nature therapy her semiretirement vocation.

Forest bathing, also known as shinrin-yoku, originated in 1982 when Japan's Ministry of Agriculture, Forestry, and Fisheries observed that office workers felt better when they were outdoors in a natural environment.

Consequently, the department started recommending nature immersion to its employees. Researchers found that after employees spent time in forests, their blood pressure decreased; their number of killer cells, a type of white blood cell that attacks tumors or cells infected with a virus, increased; and their immune system became stronger.

International research followed. In 2019, Ye Wen of Jiangxi Agricultural University in China coauthored a review of papers assessing the health benefits of forest bathing. The study found these potential benefits: "remarkably improving cardiovascular function, hemodynamic indexes, neuroendocrine indexes, metabolic indexes, immunity and inflammatory indexes, antioxidant indexes, and electrophysiological indexes; significantly enhancing people's emotional state, attitude, and feelings towards things, physical and psychological recovery, and adaptive behaviors; and obvious alleviation of anxiety and depression."[30]

Why so many benefits? Researchers offered one explanation: Some tree species emit phytoncides — antimicrobial organic compounds that serve as insecticides. (The word *phytoncide,* coined in 1928 by Boris P. Tokin, a Russian biochemist, means "exterminated by the plant.") Trees bathe themselves in these chemicals. And we get the benefit of phytoncides when we walk through trees, especially conifers.

Today, shinrin-yoku is one of several emerging therapies that utilize nature immersion. These include equine therapy, wilderness and adventure therapy, and other applications that fall under the still-evolving umbrella of ecopsychology. In Japan, trained forest bathing guides often incorporate yoga and meditation into their forest bathing practices. Janice, however, takes a more straightforward approach.

Standing up, she offers what she calls her first invitation. She asks us to taste a pine needle, close our eyes, and listen to the wind, the shifting leaves, an insistent bird nearby, and other birds farther away. "Does the bird's song have a rhythm to it? Can you hear a rhythm in the wind?" she asks. "Put your hand on your heart, feel it beating. Now, with your eyes closed, sit in the direction your body will want to sit in, and be still and listen."

I mainly hear my breathing.

After a few moments, she says, "Now open your eyes as if you are seeing nature for the first time, like a baby. What do you see? What do you notice?"

One woman, a healthcare worker, can still taste the pine needles she has placed in her mouth just minutes before. Another woman, an artist, describes the feeling of being a very young child, riding in a shopping cart with her mother pushing, and enjoying all the new sights until becoming overwhelmed by what she calls a "storm of noticing." She smiles. "And then I dissolved into a crying fit, like an overstimulated baby."

A storm of noticing. I love that phrase.

DOWN THE SLOPE, a car pulls in under the trees. A man and a woman get out. They're talking loudly.

Now it is my turn to report. I tell how I noticed a slate-colored flat rock surrounded by leaves. Staring at it, I was transported to my grandmother's rough yard. In that memory, a fermenting pear lay on the ground next to the trunk. My legs were stretched out in front of me. I saw the leather and metal leg braces I often wore when I was four years old. I turned to watch my father in his army coat as he stamped out a little fire in the grass. Then, my gaze returned to the single Volcan rock.

After a while, the man and woman leave.

As if on cue, a member of our forest bathing group reports that she can no longer hear as well as she used to. But in the forest, she can feel the air moving across her eardrums. "What I notice mostly is that I'd like to be hearing more."

"I had to work to let go of the visitors and their talking," says Janice. "When they drove away, I noticed how everything became louder. I can see the sunlight coming through my eyelids." Opening her eyes, she saw "these big tree beings. They have been here so long, lived through so much." The outer bark of the largest tree is burned to charcoal.

For her next invitation, Janice asks us to head out into the woods and "notice what's moving. Tell us what you find."

Looking down to a few feet from the ground, I notice the bowl of an oak trunk—a hollowed-out place filled with water. The air moves little rivulets across its black surface. Ants parade along the water's edge. I look up — a contrail from a jet extends slowly across the blue between the branches. Walking now, I see shadows moving on the ground, their edges quivering. I have never seen the fingers of shadows moving until now, at least not consciously. Each shadow seems to have its own personality.

I'm surprised by how quickly I'm getting into this.

Ahead, I encounter a self-healing oak. Its outer bark is mainly charcoal, but the next layer is yellow or cream, rough and light like balsa wood. On that tree, the live outer bark is already wrapping around the scars, rolling over at the edges. Look at it long enough, and you can almost see the rounded edges slowly covering the wound. Only recently have I noticed this healing phenomenon, as if the tree's inner hand is wrapping around the wound.

A few weeks ago, I was walking in our neighborhood with some visitors. One of them was Gary Knell, who had just retired as CEO of the National Geographic Society. Appropriately, he was the first to notice this phenomenon in a tree that I had walked past many times and had not seen. About fifteen feet up the trunk of this tree, thick new bark was rolling around a deep wound — either a fire or saw wound. The thick skin of the tree was enveloping the wound, making the perfect shape of a valentine heart. But that was then. Back to the moment.

Passing another black-and-bronze stump, I notice how the shards of its bark reach up like flames. Wild strawberry leaves bounce in the breeze. Poison oak, with similarly shaped leaves, is moving in around the strawberry plants.

I come to a long depression of land with a little creek rushing through it — but this is another illusion. A closer look, and the water turns out to be a type of wild grass flowing downward in clumps with long, pointed blades. Perhaps a recent storm sent a little flood down this mountainside and shaped the grasses and soil in its image. As with the shadows, the wind moves each blade of grass, and the blades join in green currents.

In the distance, Janice plays a few notes on her ocarina, a kind of flute. This is her call to reassemble.

We again sit in the grass in a circle and report the movements we have just seen or felt. The youngest member of the group, a physical therapist, discusses how her sense of scale changed when she began examining the smallest moving things.

Janice offers the following invitation: to befriend a tree. We fan out.

I head to the fallen hulk of a giant pine. I have seen it here before and have thought, "What a shame that it was cut down." Inside the trunk, which is perhaps five feet in diameter, is darkness, a cave, and a tunnel that once led to the sky through the dead trunk's center. Next to it, the stump is filled with sawdust and spiked with holes. On the ground between the stump and the fallen tree, purple mushrooms seem to pull inward, desiccated and withdrawn.

A few feet away, a circle of wood spines marks where a larger tree died long ago. Out of this ring of spines evenly spaced, every foot or so, a young tree grows out of the tree that had been. A circle of forest, rising from the carcass of the elder; an outer world emerging from the inner. I find something comforting about this moment. In the distance, a red-tailed hawk cries.

I feel a sudden wave of peace, and something else: compassion.

A few minutes later, gathered by the ocarina, someone points to a tree, high on the mountainside. Its peculiar formation of dead branches is shaped like an eagle's nest. The eleven-year-old in me wonders what it would be like to climb that tree and sit in the palm of reaching fingers.

"At some point, this is all too much," says the artist. "Overwhelming. I need a Coke."

The physical therapist speaks next. I lean forward to hear her soft voice: "See that tree over there, the big one that fell but was caught in the branches of other trees? I felt that something terrible had happened there. I sat down in front of it and asked it to tell me a story. It said, 'Others fell, but not all survived. Some of us find a way to survive.'" She pauses. The group is quiet. I wonder, as perhaps you would, what happened to her long ago and how she survived.

"I see these trees of different sizes," says Janice, "growing up together, almost embracing, holding on, and yet still individuals. Holding on to each other." She describes a favorite tree, one she has visited in this grove before. A very old tree also scarred black, the outer bark split open, the inner bark pink. The smaller pines behind it stand healthy, unburned. Janice says the elder tree probably served as a fire barrier, protecting the smaller trees from the flames.

"It made me think," she says, "we all get scared, but we can run. These trees can't. They stand and take it." Then she adds, "Sitting under that tree, I thought about getting older, how my hair is turning silver." She smiles when she says this. It is not a bad feeling, getting older.

Others also report what they saw, heard, or sensed: a broken white eggshell, large enough to be the egg of an eagle. A black ant carrying a piece of white fluff, followed by a half dozen ants that seem unburdened by any load. On a flat boulder, the remains of a mouse — its vertebrae here, its brain case over there, and most of its tiny bones clustered in one spot, surrounded by a ring of old, wet, dissolving fur, perhaps what was once an owl pellet. A feather, a leaf, each a potential carrier of seeds over a field or across miles.

One of the women paraphrases a quote she heard the other day: "If you want to get anywhere in life, you need to learn to fly."

TWO ELEMENTS OF OUR CONVERSATIONS were particularly striking. One was the tendency to turn solid objects into subjects of wonder. In my case, the stump with spires that seemed to become fire, and the wild grass, flattened by wind, that I mistook for flowing water.

What I didn't see were faces. Pareidolia is the tendency to see faces in various objects, including elements of nature — the bark of trees, clouds, or, for that matter, a cheese sandwich. In 2004, such a sandwich, reportedly marked with the image of the Virgin Mary, sold for $28,000 on eBay. The sandwich was ten years old.[31] I'm a pareidoliast (if that's not a word, it should be), as I can't pass a tree or see a cloud without that tree or cloud seeming to look back at me.

What also surprised me was how the stories — our descriptions of what we saw, heard, and felt — often sounded like parables or folklore. The child who dissolved into the storm of noticing. The "big tree beings" that had been here so long. The ants that were both leaders and followers. The urge to climb a tree and sit in the palm of its reaching fingers. The outer tree cared for by the inner tree, like a gentle nurse. And the circle of new forest rising from decay.

I wondered, Where do fairy tales come from? Or religious visions? *Is the forest enchanted?* Or have we suspended our disbelief, dismissed the pretense that we know all the answers? That's one way to describe awe.

The final ocarina notes drift through the trees. We gather for a picnic. Janice pours cups of tea she has brewed from pine needles and chickweed. She hands us the cups and pours the tea from one cup into the leaves and needles on the ground. She says, "We always give the earth the first cup."

Some of us resist such rituals. Still, the feeling of peace and compassion that filled me near the circle of new trees — that feeling was growing. It would last into the next day. The phytoncides had done their job.

I suspect the others felt this, too.

CHAPTER 6

Spellbound

T HE EARTH TELLS us stories. Sometimes it tells us our story. Other times, we impose our own.

In *William Blake vs. the World,* biographer John Higgs discusses Blake's take on perception: "The Middle English word 'vision' originally referred to a supernatural apparition," writes Higgs, "but its meaning has since been downgraded to describe regular, everyday sight."[32]

Blake, an extraordinary artist, poet, and visionary born in 1757, personified the "higher states of consciousness reported by mystics, saints, poets, and psychedelic pioneers that can be triggered by exercising the imagination."

Blake, to whom nature was "imagination itself," used what he called his "twofold vision," says Higgs. "To see the real world with his 'outward' eye and an imagined counterpoint with his 'inward' or mind's eye is, for Blake, his regular mode of perception . . . [I]t kept him on the right side of sanity."

How does a person apply this thinking to the natural world?

In *Our Wild Calling,* I report on biopsychologist Gordon Burghardt's practice of critical anthropomorphism, which he defines as "applying critical scientific thinking and imaginative empathy." Through his method of observational science, he closes the gap between humans and other animals. To study a snake, he advises, one must *become* the snake.

The process requires two steps. First, while observing the snake, bring into your mind all the hard science that you already know about snakes: What information is the snake's flicking tongue seeking? What influence does the air temperature have? And so forth.

Burghardt's second step is one that Blake would have approved of: Use your imagination. Sit with the snake; to the best of your limited human ability, imagine what the snake's skin might be feeling, what its eyes are seeing. Is it hungry? What does it sense from you? Lose yourself to become the snake. Or the deer. Or the raptor. According to Burghardt, using both modes of seeing helps students and scientists ask better scientific questions.

Such thinking would be familiar to many Indigenous peoples. Albert Marshall, an eighty-four-year-old Mi'kmaw elder in Canada, coauthored, with Louise Zimanyi, *Walking Together,* a children's book about how to practice etuaptmumk (which means "two-eyed seeing" in the Mi'kmaw language).[33] Marshall describes two-eyed seeing as viewing the environment through human eyes as well as from the perspective of animals or plants. "With my two eyes I see everything from my Aboriginal lens, and this is what governs how I should coexist in this wonderful creation of ours," he told the CBC. "In order to do that, we have to invoke the other lens and constantly seek other perspectives to learn."[34] In addition to writing the book, Marshall designed a twelfth-grade environmental studies course based on etuaptmumk.

His approach appears to align with critical anthropomorphism and Blake's concept of a twofold vision. Of his first and second ways of seeing and interpreting the world—scientifically and with imagination (empathy)—Blake wrote, "The tree which moves some to tears of joy is in the eyes of others only a green thing that stands in the way. Some see Nature all ridicule and deformity…and some scarce see Nature at all."[35]

Blake introduced two more approaches. His third perspective was to step back from focusing on a single tree and instead see it as part of a larger system. Today, while stepping back, we might "see" the mycelium networks linking trees across a forest, along with the systems of other plants and animals, weather patterns, and more.

The fourth way of seeing, according to Blake, is to back away until the individual tree and its systems fade from the mind's eye, allowing a sense of bliss to emerge. He called this fourth form of vision eternity, a "full-blown mystical state," as his biographer Higgs describes it, "one that is rare and beyond words and self."

These perspectives on the world are what Blake ultimately called "fourfold vision."

WHEN BLAKE DISCUSSED vision, he likely wasn't just talking about sight but all our ways of perceiving. Higgs believes that neurologists are only now beginning to understand how imagination works and that "it is very different from person to person." Higgs continues: "If you ask several people to close their eyes and imagine a cow, the results will vary greatly… Some people will see nothing at all, which indicates a condition known as aphantasia, experienced by individuals who have little or no ability to create mental images."

Neurologist Adam Zeman first coined the term *aphantasia* in a 2015 paper. Aphantasia exists on a spectrum, with hyperphantasia at the other extreme end. Hyperphantasics, in Higgs's interpretation, "tend to spend more time daydreaming than the average person" and are more prone to "regret, longing and nostalgia; they experience both greater anxiety and greater empathy than most."

Higgs theorizes that Blake was a hyperphantasic.

"If you could glance inside a hyperphantasic's mind when you asked them to imagine a cow, what you saw might shock you," Higgs writes. "A hyperphantasic would see a cow in their mind's eye as vivid and real as if it was standing just in front of them. They would not only be able to see the cow in pin-sharp focus, but they would also be able to smell its damp hide and feel the heat from its breath."[36]

BOTH SCIENCE AND INDIGENOUS knowledges suggest we have the potential to see more deeply into nature, but how often and how clearly? (Not long ago, I saw a man wearing a T-shirt that read "My third eye needs contacts.") A bit of skepticism is in order.

The writer and Santa Fe Institute scholar Laurence Gonzales offers this cautionary tale.

As a boy in San Antonio, Texas, Gonzales loved a ceramic ashtray his grandmother had bought in Mexico long ago. The ashtray was made in the shape of a coiled rattlesnake. "It was dark and dusty and very realistic," he recalls in his 2008 book, *Everyday Survival: Why Smart People Do Stupid Things*. "As I grew up, I always wondered what had become of that ashtray."[37]

Years later, as he was hiking, "I came across the ruin of a stone house," he writes. "As I poked through the wreckage, suddenly there was my grandmother's ashtray, complete and unbroken. Delighted, I reached out to take it. And then its tongue came out. I froze. Its tongue came out again. The serpent was smelling me. With the hair standing up on my neck, I carefully backed away."

Thinking later about this brush with danger, he recalled his emotional connection to his grandmother's ashtray. It had become "a strong, persistent model" of a rattlesnake that could not hurt him. "I was operating as if under a spell."

The spell of the senses that some of us experienced in the woods when we were growing up still helps us interpret the world. But not every spell cast by nature is beneficial.

As infants, our brains create mental models of everything we see and otherwise experience. When the brain receives the first bit of information about what it's looking at, it substitutes the mental model "because that's the most efficient thing to do," according to Gonzales. And at times, the safest. "Your brain does not want you to see the world [as it is]. That's too much work. You'll be eaten by the tiger."

Gonzales was endangered that day by his old mental model. He was also operating, he surmised later, with a lack of curiosity about the real world. The good news is that the experience strengthened his chances of survival in the coming years by encouraging him to be more cautious about what his senses tell him about the story he thinks nature is telling.

One approach to getting nature's story right is taught by the botanist Doug Tallamy, author of *Nature's Best Hope*, a book about reviving native species, mainly plants. Tallamy is also a professor in the Department of Entomology and Wildlife Ecology at the University of Delaware. He advises his students to look for what he calls the *search image*.

"Studying a book or doing a Google search for images will help," he says. "But you'll probably have a hard time finding what you're looking for until you encounter it in its natural habitat." Context counts. "Neurologically, you must see what you're looking for before you can search for it. You must see a mountain lion track before you can see a mountain lion track." Here's an analogy: When you first drop a contact lens on a rug, you can search and search and not find it. However, you might then see a glint of light from the lens. As a result, the second time you drop the lens, you'll recognize the glint, which is the search image. It leads you to the rest of the contact lens.

Tallamy recounts how one of his fellow professors hired undergraduates every summer to search for nymphs of treehoppers. (A treehopper is an insect related to the cicada but is more flamboyant. It can customize its body parts into tiny helmets shaped like horns. Treehopper nymphs sometimes look like snowflakes with legs.) Tallamy's professor friend would send the students out to look for the nymphs, even suggesting which trees the nymphs frequent. The students often returned baffled, with no nymphs.

Then the professor would take the students back to the same trees and point out the treehopper nymphs that had been right in front of the students' eyes. Once the students saw one of the strange nymphs or part of one, they had a search image. Suddenly, nymphs seemed to be everywhere.

I asked Tallamy if such skills of noticing were taught at his university or any other university. "No, not as a course," he said. "But I do teach my students about search images. I wedge that into the curriculum." Maybe someday, he could broaden it into a course: Noticing 101.

FROM A YOUNG AGE, our prehistoric ancestors developed hundreds of noticing skills, and they didn't wait for all the research to come in. They just wanted to avoid being brunch for a bear. Or to be able to spot a poisonous viper before the viper saw them.

They probably used senses most of us don't know we have. Forget the traditional limit of five senses — touch, sight, hearing, smell, and taste. Depending on which researcher you ask, the number of recognized human senses now ranges anywhere from nine to thirty or more, according to scientists studying the human senses.

We use some of these senses without even knowing it; others we fail to develop.

Aristotle compiled the original list. He posited that for each sense, there is a specialized sense organ. These were the eyes, ears, nose, tongue, and skin. Two millennia later, science added the vestibular system, located in the inner ear. That system is for body awareness, which, in addition to helping us stay physically balanced, allows us to focus our two eyes on a single point when we're turning our heads. As it turns out, we don't need a sense organ to know which way the wind blows. We have specialized cells called sensory receptors, which are located primarily in the skin, eyes, and ears.

To the traditional five senses, John M. Henshaw, a professor at the University of Tulsa and author of *A Tour of the Senses,* adds four more: balance, temperature, pain, and proprioception, which is the awareness of the body in space. Proprioception relies on joint position and joint motion — senses that respond to stresses placed on the body when we move. We see with our eyes but also with our body. We *feel* what is required to move through our surroundings. How we see the world is linked both to our visual acuity and to the threat we might sense from the corners of our eyes. Unconsciously, we calculate where we might turn or fall, and in what direction. We're a blink away from what happens next.

To the list of senses, also add chronoception, the subjective experience of time passing.

If, in addition to identifying previously unknown or unappreciated "new" senses, we mix and match the standard five, then the question of how many senses we have grows more complicated. For example, some people without typical sight can learn to sense their location not only by locating sounds, such as the sound of leaves moving or water flowing, but also by understanding how sounds bounce off surfaces.

Surely some of our supersenses help us stay and feel more alive.

Recent research shows we have some of a bloodhound's ability to follow a trail by scent. For a UC Berkeley study, researchers recruited college students who participated with their eyes blindfolded and their ears plugged. On all fours, they followed a trail of chocolate through a forest. The researchers noticed that the students often zigzagged, a habit shared by animals that make their living with their noses. Zigzagging helps them avoid losing the trail.

Making sense of the senses is complicated not only by the newly discovered sense outliers but also by how multiple senses, including the basic five, team up.

Synesthesia is a neurological condition in which stimulation of one sensory or cognitive pathway triggers the activation of another. Siri Carpenter, writing in the American Psychological Association's *Monitor on Psychology*, reports, "Some synesthetes hear, smell, taste or feel pain in color. Others taste shapes, and still others perceive written digits, letters and words in color. Some, who possess what researchers call 'conceptual synesthesia,' see abstract concepts, such as units of time or mathematical operations, as shapes projected either internally or in the space around them."[38]

She adds, "About one in 2,000 people are synesthetes, and some experts suspect that as many as one in 300 people have some variation of the condition. The writer Vladimir Nabokov was reputedly a synesthete...and the physicist Richard Feynman."

Is synesthesia a bug in the neurological works or a feature?

Early hunters and gatherers may have developed, mastered, and celebrated such sensory abilities for their practical advantages and for their spiritual experience. Could this be why so many people (at least in my interviews) describe their encounters or relationships with other animals or even plants as altered senses: a shift in the sense of scale, or the impression that time is bending?

In a sense, we're blessed with more senses than we know what to do with. This poses a challenge to education, and not only at the college level. It's an opportunity. What self-respecting sixth grader wouldn't want to learn about their supersenses? And we haven't even talked about ESP and dermo-optical perception, the alleged ability to see with fingertips — strangely akin to the belief of some botanists that cells at the root tips of several plant species allow the plant to "see" obstacles as its roots make their way through the substrate.

Henshaw cautions that it's more difficult than it might appear "to put a definitive figure on the total number of senses that humans possess. At some point, it becomes just a bit arbitrary."[39] The good news: We can get the big picture without all the details. The not-so-good news: We can mistake a rattlesnake for an ashtray.

NEUROLOGICALLY AND THROUGH cultural habit, perception of reality depends on how, where, and when we pay attention.

In the early 1970s, the husband-and-wife research team Stephen Kaplan and Rachel Kaplan developed what they called *attention restoration theory*. The Kaplans were environmental psychologists inspired by the work of psychologist and philosopher William James, who, in 1890, identified two kinds of attention: *directed attention* and *soft fascination*.

In a nine-year study, they followed participants in an Outward Bound–like wilderness program that took people into the wilderness for up to two weeks. The Kaplans reported that during and after their long hikes, the participants reported a sense of peace and an ability to think more clearly. Just being in nature felt more restorative than more strenuous experiences, such as rock climbing. The Kaplans were surprised at the benefits of what they came to call the *restorative environment*.

Their research revealed that excessive directed attention (such as staring at a computer screen or driving in urban traffic) leads to what they call *directed attention fatigue*. Symptoms include agitation, irritability, difficulty concentrating, and impulsive behavior. This form of fatigue (or burnout) is produced when neural inhibitory mechanisms spend too much energy blocking competing stimuli. Quoted in *Monitor on Psychology*,

Stephen Kaplan explains, "If you can find an environment where the attention is automatic, you allow directed attention to rest. And that means an environment that's strong on fascination."[40]

Therefore, time spent in the natural world can be the most effective stimulant of restorative attention. One explanation is that natural environments are often distanced from stressful homes and workplaces. But you don't have to head for the hills to get the relief you need. Rachel Kaplan also found that office workers who had a view of green space exhibited some of the same relief. In a greener office, workers are healthier, report greater life satisfaction, and (good news for employers) like their jobs more.

Suzanne Gaskins, psychology professor emerita at Northeastern Illinois University in Chicago, describes *open attention* as a form of meditation practiced in some rural cultures "that remain more attached to the natural world." Both forms of attention, focused and open, are required to know the world. To develop focused attention, you train your mind to reduce distractions and resist the messiness and chaos of the world. By contrast, open observational skills are relaxed, alert, and wide-angled — and probably harder to teach.

With open attention, you don't resist chaos. You observe it and withhold judgment, to the extent that you can. You just take it all in.

While recording her teaching process, filmmaker Carol Black noted a distinct change in students after they learned about and practiced open attention. "They noticed everything," she says. "They remembered everything." Adults accustomed to taking students on group outings "would say they had never seen kids like this before... They expected all children to be wound up, tuned out, half-frantic with suppressed energy, like a dog who's been locked in the house all day."[41]

Similarly, the primatologist Frans de Waal advised people to use what he called *holistic observation* of nature: "Looking beyond the central scene is key."[42]

Scientific inquiry usually begins with an act of the imagination. A conjecture becomes a hypothesis, which in turn seeks a conclusion. At its core, science steps beyond the known; it forms a search image and examines it, and sometimes the envisioned turns out to be real.

Otherworlds exist all around us. Most might appear to be ground fog, faint singing, or fleeting shadows moving through or above the trees. These worlds emerge from under the bark, from wind and clouds, from deep beneath our feet, but cannot be seen by eyes alone. The shapes, sounds, and perceptions are created by biological life and geological forces, or they may be imagined or projected from within us.

These otherworlds — the mycelium messages whispering through forests, the dance of predators and prey, the geopolitics of ants, the electrical fields of plants and bees, the nighttime biosphere of bioluminescence, our own glittering bodies — these and other realms have always existed. So easily dismissed as superstition, the feelings of awe they produce are captured in stories and myths inspired by living, breathing creatures; wind, rain, and snow; floating seeds. And falling embers.

TO NOTICE OR NOT to notice is a choice, and the choice depends on context.

By making that choice, the late Jane Goodall changed the way we view the world. The primatologist, anthropologist, and foremost expert on chimpanzees, Goodall was only twenty-six when she went to Tanzania in 1960. At that time, under the sponsorship of Louis Leakey, she had no degree or academic standing. The traditional way, at least scientifically, to observe wildlife would have been to measure and quantify. However, Goodall sat on a mountainside with only a notebook, binoculars, and her fascination and simply watched. Week after week, month after month. Naming the chimpanzees and identifying with them to the extent she did was then anathema to traditional science, or at least unconventional. But Goodall identified them by names like Fifi and David Greybeard and, over the years, came to know them deeply.

Through her observations and empathy, Goodall's work has changed how we view chimpanzees and many other living creatures. She witnessed a range of humanlike behaviors, from kindness to armed conflict.

Reporting her research, she challenged the long-standing beliefs that chimpanzees were vegetarians and always peaceful and that only humans could make and use tools. These findings were revolutionary. Leakey wrote later, "Now we must redefine 'tool,' redefine 'man,' or accept chimpanzees as human!"

When I spoke with her at a 2023 conference organized by the Children & Nature Network, she offered this thought: "As people, we have a spoken language, and we like naming things; we have to have a word. So, we call the connection we have to the natural world, to the spirit world, whatever it is, we call that the soul. Our soul. So, if I have a soul, then a chimpanzee has a soul as well. All living creatures have the spark of the divine. The Indigenous people have long understood that the animals and the plants and even the rocks and the mountains are all connected to us in a spiritual way; they are all our brothers and our sisters."

In 2010, Oliver Sacks, the British neurologist and naturalist who died in 2015, wrote about the cognitive disorder prosopagnosia, which hinders a person's ability to recognize familiar faces, even including one's own.[iv] He quoted Goodall, who said, "I have huge problems with people with 'average' faces. I have to search for a mole or something. I find it very embarrassing! I can be all day with someone and not know them the next day."[43]

Occasionally, recognizing places was also difficult for her. "I just don't know where I am until I am very familiar with the route. I must turn and look at landmarks so I can find my way back. This was a problem in the forest, and I often got lost."

Perhaps her so-called limitation made her a better observer and more vulnerable to nature's embrace.

---

iv   Goodall also acknowledged that she had a mild case of prosopagnosia, a difficulty recognizing faces. Brad Pitt reportedly shares this condition.

CHAPTER 7

Backyard Wonder

Down the mountain, our friends Bill Stothers, a retired journalist, and the Reverend Cyndi Jones, a Lutheran minister, live in San Diego in a densely populated urban neighborhood called South Park. Bill, a former editor and ombudsman at the *San Diego Union-Tribune*, contracted polio as a young boy. So did Cyndi when she was a toddler. She spent months in what was then called an iron lung. From 1982 to 1998, the couple published *Mainstream*, an international advocacy journal for people with disabilities.

Today, rolling along the sidewalks of San Diego, Bill points his camera at what would be waist or knee level for people capable of walking. His lens is often pointed up at street signs, the limbs of trees, graffiti, the wonder of light and sky.

Cyndi uses an electric scooter. Her perspective is only a few inches higher than Bill's, but she says he notices things in the neighborhood that she does not. Her eyes, she says, are instead particularly attuned to the unnatural, the out of place, to danger and oddities. I ask her if she usually sees such things indirectly, through her peripheral vision.

"Come to think of it, I do," she says.

She tells me about something extraordinary she noticed in their back-yard: "a gift," as she calls it, or something imagined. Or both.

She caught sight of it when she was looking in her refrigerator. Reaching to grab a milk carton, she glimpsed in her side vision a bouquet of flowers in the backyard. The flowers were momentarily lit by the setting sun.

The oddity was that it was sticking out of the bark of the palm tree—which doesn't have bark, in fact. Instead, it has sclerified (hardened) cells left behind when fronds fall from the tree. She showed me a photo of the bouquet. High on the trunk was a cluster of white flowers, finished with long green leaves embracing the flowers. For all the world, what she saw looked like a professionally arranged bouquet.

"I thought, Who would place such a thing in such an unlikely place?" she says. "Was it a gift from someone, a neighbor who liked us?"

The flowers were in fact an inflorescence, an arrangement of flowers on stems that appeared to comprise a bouquet.

Cyndi quickly realized that it had not been placed there, that it was exactly where it (rarely) belonged. But as a minister, she chose to interpret the bouquet as a spiritual message gifted, or regifted, to her and to Bill. Fifteen years earlier, the yard had been dirt, apart from one weed, which, for some reason, Cyndi had decided not to pull up. The weed had grown into that palm tree. "People say this was coincidence," she says. "Well, maybe." Then she adds, "God is in the timing."

SIGNS AND WONDERS. This morning, walking on the road that loops higher into Pine Hills, I feel light and good about the world. And then the land grows darker. I follow the truck tracks and deer tracks and the fading light in the trees, down a dip, into a melancholy place where I do not think the kindest, best thoughts. Some places are like that. But I keep walking.

Seeing faces in the bark of trees. Hearing voices in the wind. Fairy rings. In contemporary Western culture, though not in many others, such impressions are often discounted as self-delusion, hallucination, or superstition. I prefer to think that these visionary moments grow as we understand science and as we connect with our deepest need to find some meaning in life. A tree's bouquet, a face in the clouds, or the mesmerizing stare of a mule deer can be a projection. Or a truth.

Is this so strange? We also find meaning in works of fiction and works of visual art—all ephemeral, all projections. We discover meaning in the words and brushstrokes, but we also find it between the lines. Ernest Hemingway advised writers to leave space for the creative reader to enter. Nature's art offers us the same opportunity for mysterious stories that we sense or hear and see in a new way.

Scientific inquiry depends on imagination; the process always begins with a question, a ventured hypothesis first imagined. A biologist looks through the lens of a microscope and is overcome by wonder. Humbled by nature's complexity, seen and unseen, some scientists designate themselves *mysterians*. According to the multidisciplinary journal *Nature Neuroscience,* mysterianism "holds that even though there is nothing supernatural about how consciousness arises from neural activity, the human brain is simply not equipped to understand it."[44]

Since moving to these mountains, I have adjusted my view of dark

places, which may or may not exist, replacing that view with the old Celtic belief in thin places, or locations where the physical earth we know is separated from heaven by only three feet, give or take.

If not in the physical world, thin places do exist in the imagination—or some other consciousness we share but barely know. *Thin place* is an "odd term," writes Eric Weiner, author of *The Geography of Bliss.* "One could be forgiven for thinking that thin places describe skinny nations (see Chile) or perhaps cities populated by thin people (see Los Angeles). No, thin places are much deeper than that. They are locales where…we're able to catch glimpses of the divine, or the transcendent or, as I like to think of it, the Infinite Whatever."[45]

I'm partial to that term: "the Infinite Whatever."

The Northern Irish writer Kerri ní Dochartaigh is a bit more reverent and poetic about thin places. In her book, called just that, *Thin Places*, she describes them as "places that make us feel something larger than ourselves, as though we are held in a place between worlds, beyond experience…Some seem to call us back to them again and again." These places "dance on the caves of our insides, even as we try to cover them up from view."[46]

I LOOK FOR THOSE PLACES now. In Julian, there's a dusty truck trail that snakes through the hills above a little lake. It's not an official hiking place, though from time to time, people do park along the two-lane highway and duck under the gate, often with their dogs. When I walk this road, it feels like a tunnel, a passageway.

Once, my hiking buddy John and I were walking this trail of dust in the hot sun (he and I agree that the best hiking weather is what most people consider bad weather). Ahead of us, a long Southern Pacific rattlesnake was slowly making its way across the path. We stopped and watched it disappear into the vegetation.

Then we looked around. We were surrounded by snake tracks sketched into the dust over time. Above us was a jagged outcropping with a jumble of sharp boulders, and between the boulders were good, dark places, the mouths of minicaves. Standing there and looking up at those rocks, I thought of the ravine of winding vines from my childhood. Near that ravine was an irresistible old well filled with mud, fallen leaves, and other debris. I descended into that well at age thirteen, in search of snakes. And I found them — a nest of copperheads! And instantly learned to *dance*. I pumped my legs and waved my hands like Michael Jackson, then sprang from the well as if shot from a cannon. That's when I developed my phobia of…dancing.

Though thin, these were not dark places. In fact, today, they light up in my memory.

Up the road from us lives my eighty-nine-year-old friend, a former fire chief and still our neighborhood's unofficial firefighter, Jim Baker. He has his own theory about thin places. One day, while we were talking under the oaks, he said (without knowing I'd been researching thin places), "You know, heaven and hell are very close, so close that if you don't look up, you'll be in hell. But all you have to do is look up and look around and you'll be in heaven. Look at *all this.*"

Another friend of Volcan Mountain, the late Bill Carter, lived off the grid for decades. While I tend to consider place horizontally, Bill thought vertically. As he gazed with wonder through the lenses of his telescopes, his sense of place extended from the earth below his feet through a column that reached other constellations, ones that exist now or did so billions of years ago. In Bill's column of place, no one species or star is immortal, and none matters more than the others.

His vertical view gave him a different kind of hope. If the destructive-ness of our species leads to our own eradication, other life will emerge, other intelligences — perhaps keener than ours, here or among the outer stars.

A sense of place can emanate from animals and plants and geology and weather, or in some cultures, place can be perceived as spirits. To ancient Celts or Indigenous peoples, a place itself had its own senses. Others con-sider place to be literally a *sense,* a combination of smell and taste or a kind of sightedness. Or a sense of place is a projection: the sum of the attributes that we detect in a particular place or project onto it.

To enter that world requires both imagination and science.

The writer Gertrude Stein once said of her hometown of Oakland, Cal-ifornia, "There is no there there." Of the natural world, I wonder if she might have said, "There's *more* there there."

CHAPTER 8

Virtual Realities

I N THE FALL of 2022, I was asked to participate in a conference panel discussion sponsored by a New York museum. Four academics and an artist would speak from the stage. I would participate remotely, through Zoom. The purpose of the conference was to ask this question: How can the metaverse connect us to the natural world?

The term *metaverse* refers to a networked virtual world in which avatars — images representing people or other life-forms — interact, do business, and play games usually in 3D. For example, in an online virtual world called Second Life, people (through their avatars) can develop relationships, have virtual sex, and even get married. A metaverse can also be used as a virtual platform for creating "digital twins" of physical spaces or objects for simulation and design.

Could the metaverse be extended to real animals and plants? Wildlife conservationist Gautam Shah and novelist Ryan Boudinot think so. As they explain in the *Stanford Social Innovation Review,* "The data to create unique, digital identities for thousands of animals is at hand." Currently, scientists and researchers are tracking more than 200,000 individual animals by GPS, and more than 800 species via satellite. These plants and animals could theoretically be assigned their own avatars.[47] If all that data were poured into a virtual world, people would have the chance to get to know the digital twins of individual animals.

Through a tree avatar, for example, a person might "experience" real-world deforestation in a deeper, more personal way. Or they might get to know a specific orangutan named Fio "hanging from a fig tree," and notice that "she's looking a little skinnier than the last time you saw her." Later, "a world-famous orangutan researcher from Malaysia sits down, and you strike up a conversation about Fio and the state of the Bornean rainforest." Such experiences, unreachable to most people in the physical world, might lead to a more personal relationship with nature. Conservation, then, becomes less abstract and more personal.

Most of the onstage participants at the New York conference were enthusiastic about the potential for the metaverse to bring wildlife and humans closer. One panel member argued that the human experience of biological nature is already migrating to virtual reality through artificial intelligence, gaming, and the internet. Another panelist suggested that blockchain, the vast information storage and conveyance system that uses millions of computers (possibly including the one on your desk), is every bit — or byte — as much a part of nature as a biological forest — functionally similar to the fungal network that allows trees to communicate. Anyway, the panelist added, people who never go outside won't miss real-time nature all that much. Better to experience virtual nature than no nature at all.

Indeed, recent research has shown that people experiencing techno-logical nature can, in fact, receive some of the stress reduction and other benefits of true nature connection.[48]

John Mack, an artist and photographer, made a counterargument. His work, which includes an educational forum titled "A Species Between Two Worlds," challenges the rise of artificial intelligence and digital immersion. He asked, "Will we decide to drop our humanity by the wayside, or will we ensure its survival into the future?"[49]

The human migration from the real to the virtual began long before the personal computer was invented. From the beginning, according to Mack, humans have mistaken the *image* of nature — the painting on the cave wall, later the photographs and pastoral art, and now virtual reality — for nature itself. As Mack sees it, we are born with the supreme skill of visualization, but the moment we interpret nature through a human-created image, we become distanced from its reality.

What about photographers, musicians, and other artists who share their interpretations of nature? Surely, artistic expression can also bring us closer to nature. But if nature is perceived only as scenery, a good view, a potential postcard, a gamer's paradise, then nature exists not for its own sake but for ours, as visual pleasure.

In a 2021 interview with Kara Swisher in the *New York Times,* author Jeanette Winterson offered this aside: "And you know, in China, there's a group of people who call themselves two-dimensional, because they say that their significant life at work and at leisure is on the screen, and that their body, in fact, doesn't matter at all and they're not interested in it."[50]

Coincidentally, this conference took place around the time that Apple announced its Vision Pro mixed-reality headset. A reporter for *CNN Business* tried it out and reported her experience: "In the demo…a virtual butterfly landed on my finger; a dinosaur with detailed scales tried to bite me…When a small bear cub swam by me on a quiet lake during another immersive video, it felt so real that it reminded me of an experience with a loved one who recently passed away. I couldn't wipe the tears inside my headset."[51]

Eventually, the conference's monitoring camera turned from the live bodies on the stage to me, sitting in my office, looking at the hall through Zoom. They asked what I thought about technology's ability to sub for nature. Great, right? This would be interesting.

"I don't think so," I said.

Simultaneously, for the first time since our conversation began, I realized that my face was projected onto a big screen visible to everyone in the auditorium. The attendees had witnessed every flinch, every knuckle poke to my head, every Olympic-level eye roll I'd made.

Busted.

*　*　*

I'D BEEN THE ONLY virtual presenter. And here I was, being all judgmental.

After the conference ended, I felt like a digital shadow of myself—an avatar. So, I walked away from my computer and headed outdoors.

A walk under the real sun calmed me down.

I saw a group of graceful young deer watching from under the trees. They were beautiful. I stopped to stare at a slab of granite. I had often passed this rock without noticing the carved holes in the smooth surface. This time, I studied the holes. They were about eight inches across and a foot deep. Their sides were smooth, and at the bottom of each hole were green flakes, bits of yellow grass, fallen leaves, and patches of dark-olive moss or lichen.

A hundred or a thousand years ago, Kumeyaay women sat on the slabs as they ground acorns and other seeds into flour.

The mortar holes were known as *ehmuu* in the Kumeyaay language. Such sites are not uncommon in these mountains and are usually located on a ledge or an outcropping of granite.

Nearly always, the grinding rocks overlook a defensible space like the one I am looking at now, the long valley beneath Volcan Mountain, the peak that the Kumeyaay called Hahachepahg. This is when imagination, the metaverse within each of us, kicked in.

Now, with a shift of mind, I see snow-capped Hahachepahg when it was ten thousand feet tall, twice its current height. I can see the flaming rocks flung across the mountain by a volcano on Mexico's mainland. The peaks rise and fall in time. The land breathes, dissolves, and rises again. Fire consumes the valley.

A forest appears. I watch ancient animals move below, and the first human hunters slipping through the trees. The women are singing and talking about their lives and the lives of others. Their eyes are drawn across the valley to where the grass is yellow and the sun is high. The women lean into their work. They hear the crunch of acorn shells, the call of a red-tailed hawk, the wind picking up.

This is a different way of noticing what is real, what is not, or what is between. We exist not only on a landscape but in a timescape. A dream-scape. And never alone.

One of the women on the rock leans back and feels the breeze on her face, and inhales the scents of cedar, pine, and oak drifting up the valley. A Pacific fog fills the valley, and Hahachepahg becomes an island on another sea.

Now I hear the local donkey named Huckleberry braying and setting the dogs barking on the next hill over. I stand there listening as the womens' singing returns to the wind.

YOU CAN FIND other messages from across time throughout these mountains and in the desert below: Indigenous pictographs of humans with impossibly stretched torsos. Faded red shapes of human and lizard skeletons. Wheels of stone laid out by human hands to track the stars, patches of granulated granite, ancient roasting pits. A yoni fertility symbol carved into a rock hollow that still catches and stores water, and another stone reshaped into a penis and testicles.

Then, down the eastern slope of this mountain ridge, a petroglyph interpreted as the Anza-Borrego Lizard Man. And in the Baja extension of this bioregion, a rock image of horned El Diablito, the little devil once drawn by counterculture cartoonist R. Crumb, who advised us to "keep on truckin'."

And so we do.

In another time long ago, you might have rested under an ancient oak in the backyard or under the tree a hundred feet from me, on this ridge. You might have felt yourself held to the sky by the tree's limbs or by the arms of a father or mother.

As Jared Farmer, author of *Elderflora: A Modern History of Ancient*

Trees, puts it, such a tree can be "a bridge between temporalities we feel and those we can only think."[52]

We are not the only time travelers. Stanley Coren, a neuropsychological researcher, argues that dogs' ability to "detect both new (strong) and old (weak) scents" allows them to perceive "events that occurred across intervals of time." By the scent of our shoes when we come home, our dogs know where we've been and possibly when. They can, he says, "smell time."[53]

If we notice, nature has the power to carry us across time.

I think of my friend Bill Carter staring into his telescope in nearby Cuyamaca Woods. With his vertical sense of place, he could imagine himself into the stars and through time. The sensation of such a moment might be replicable by a computer, but a flat panel with Dolby speakers cannot duplicate the feelings he must have experienced. The touch of damp or dry air, the faraway coyote singing, the unpredictable starlit cloud approaching from the west. Some of that experience can be found on Google Earth or in virtual reality goggles. But an alarm would sound somewhere in the mind: This is not real. The only sure way to be consumed by the stars is to sit or stand under them, then look up. Like Bill, that night.

Human physiology — eyes forward, legs built to walk in a linear direction — is largely responsible for how we perceive time, according to one theory. Unlike us, an octopus moves in all directions — no forward, no backward — and probably experiences time very differently than we do. How differently? "Current research methods may be able to take us only part of the way toward an answer," says David Borkenhagen, a scholar at the University of Calgary, "but it's far enough to consider a radical possibility: if we became more like an octopus, could we free time, metaphorically speaking, from its constraints? Could we experience it as multidimensional, fluid and free?"[54]

Human cultures also determine how people perceive time. In time-studies language, some cultures take a monochronic, linear approach to time — get the business done, meet the deadline, time is money. One deadline at a time. Others are polychronic, nonlinear, or cyclical: They invest more time in building relationships and do more multitasking. In the natural world, what goes around comes around. Seasons, migration patterns, the animals that show up in our yard at certain times of the year, the deer that dies, the new fawn curled in the grass. At least in my experience, the closer we are to nature, the more time seems to move in circles.

Marcia Bjornerud, author of *Timefulness*, argues that science should be taught within the context of time and place, because "every person, idea, culture, organism, landscape, and continent exists in a particular time-stamped moment while also bearing vestiges of a much deeper historical and evolutionary past... [T]heir paths through time shaped what they are now, and what they may become in the future. In other words, all things are full of Time — they're Timeful."[55] She describes a natural world "bursting with backstories — tragedies, comedies, sprawling million-year sagas, chronicles of forgotten empires," with all these histories shaping the present. Ocean quahog clams can live five hundred years or more, and Greenland sharks take a century to mature.

"The memory of the world is deep enough to lose yourself in it," naturalist Sarah Hatfield writes. "And while we humans often choose to fill our memories with and measure the passing of time by loss, nature is more forgiving... The memory of the world (not the human world — the greater world) knows that the mistakes and triumphs are nothing more than the leaves of one autumn blown through the landscape."[56]

Could a virtual reality machine replicate any of this? Maybe someday. But right now, look over there. See the rocks and grinding holes and the bright clothes of the singing women. And down in the Holocene valley, prehistoric camels and three-toed horses, and a mastodon scraping its back along low branches. And then the First People, walking slowly through the redwoods that once were. If we can inhale land and sky in the loudness of time, can we escape the past and imagine the future? Can we look down into the valley and toward the mountain, still shrinking, and see wildfires to come, the valley barren — or fertile with revived life, with species never seen in this time, new ones migrating from the south, evolving in a newer world? Or people who might live in that future or another, sitting on this rock, singing or silent?

CHAPTER 9

When Trees Hug Back

To help us connect deeply with the natural world and our inner nature, ecotherapists begin with a few basic actions. One is to find a sit spot in a natural environment. Go there often, sit still, open your senses, and notice who or what shows up.

Some of us have more trouble than others sitting still. I practice what I call the *walking sit spot*—also known as walking meditation, a Buddhist concept that involves preferred, but not required, internal rituals involving breath, focus, and using the senses. In other words, noticing. The oaks and the hawks do the heavy lifting.

In the book *Natural Mindfulness*, Ian Banyard shares how, when he was a boy, his mother struggled with bipolar disorder. He found relief and calm through walking, especially in natural settings.[57] Walking with natural mindfulness, he says, can help "access a profound clarity of thought, an acceptance that life is what it is, a sense of peace and inner balance, and the wisdom that our true nature can be found and healed only in the present moment." Rather than use your mind to guide your senses, use your senses to guide your mind.

"The key to walking mindfully is to pay (or invest) attention to the present moment," Banyard advises. "Practice intentionally moving your focus between panoramic vistas and tiny details like the veins on a leaf or the way sunlight dances on the surface of a lake."

And walk, talk, and *think* quietly. "I remember on many walks coming across other walkers who, despite being in beautiful landscapes, constantly chattered to each other. Filling the air with thoughts, judgments, and opinions about their past experiences, or worries, concerns, and plans about a future time that was yet to come," he says.

Guilty as charged. Thinking about my hikes with pal John Johns, I realize we must sound like breaking news on CNN's *Newsroom Live*.

CARRE ST. ANDRE, a writer and former local real estate agent, tells me how she found her sit spot. Diligent, hardworking, and successful, she is dedicated to helping people find their ideal place in life. She worked with us for two years to find our current home.

Carre is a born romantic, despite a problematic history in that department.

She grew up on San Diego beaches under the aggressive male gaze, then joined the military. A former paratrooper, she made her first jump in 1988 at US Army Airborne School in Fort Benning, Georgia. She also earned three black belts in martial arts.

She moved to these mountains to marry a man who later became an abusive alcoholic.

"He'd fill the air with disruptions and demands," she says, and he was violent. "During all that time, my main refuge was to hike up a steep hill before the sun came up. My special sit spot was a rock that overlooked a waterfall in the canyon. I called it my butt rock. Everywhere I've lived, I've found a butt rock."

At the time, she claimed no specific god or any formal meditation practice. She would simply climb the hill to the rock, sit down, and silently wait for a bee to land on the flower in front of her, or a California jay to verbally claim its territory, or an opossum to waddle by, ignoring her.

"I would settle into the background like white noise. I would listen to crickets, feel the breeze on my skin, notice the subtle nutty smell of buckwheat."

Then one day, while sitting on her butt rock, it moved. "I felt the mountain underneath me shift from contraction to expansion, and I realized that mountains breathe, too."

When that happened, she began to cry. With gratitude. For her life.

A few months later, she was driving up Highway 168 toward the Ancient Bristlecone Pine Forest in the White Mountains of the eastern Sierras. "My head was full of hesitation and self-doubt," she recalls. "Why was I driving up into these remote mountains by myself? But my heart is following an obsession with old trees." The obsession started at her house, a few miles from Julian, as she watched the oaks regenerate after recent wildfires. It deepened later, during a trip to Oaxaca, Mexico, when she encountered a two-thousand-year-old Montezuma cypress.

Now, as she drove up White Mountain, the temperature dropped significantly.

The goal of her planned hike was to find Methuselah, the oldest known single tree on Earth. "I was anxious to see if I could figure out which tree was Methuselah," she recalls. She describes being humbled by the high altitude. By the time she completed the 4½-mile loop, she was no longer looking for Methuselah or any single tree. She felt like a speck "entering a vast ocean from a river mouth."

Carre finally reached Patriarch Grove and wondered why it was named Patriarch and not Matriarch. She picked a sit spot and quietly observed. "Many of the trees were ninety percent deadwood," she tells me, "with a thin ribbon of cambium delivering nutrients to one living section. There were dead trees still standing after eight thousand to nine thousand years that still supported the living."

Some trees, she says, "appeared to hold a fainted loved one in their arms. The downed trees protect the shallow root systems from erosion… The lines of life and death are blurred in the forest, dying trees sending nutrients to their seedlings. The fallen trees surrendered at the others' roots, preserving precious little rainfall, holding erosion at bay, little by little, asking nothing in return."

She watched the life that the arms of the tree supported: Clark's nutcrackers, pinyon jays, mountain chickadees, ravens, and many more. The trees have given their support for thousands of years.

"They're the oldest 'living' things," she explains, "but they're mostly dead — a slender swath of cambium, a strip of bark, carrying scarce moisture and nutrients to a burst of pine needles." And yet they care for trees yet to be born.

Sitting on this spot among the bristlecone pines, Carre was again overwhelmed with emotion. "How can something so indifferent to me, that has survived since the Pyramids of Egypt were built, and the rise and fall of many empires, elicit such a deep well of emotion?"

Since Carre shared her bristlecone pine story with me, I've often thought about her question about the unexpected emotion she felt that day. I've had a similar response. Many of us have. I've come to believe that such an emotional response to nature can help heal a deeper isolation.

Another friend tells me about how, several years ago, she was struggling with a breakup or a loss of some kind. Uncomfortable with confiding in anyone, she found herself in the little park at the edge of Julian. Wrestling to work through her torment, she became aware that she was sitting between two large trees. She was overcome then by the sense that both trees were leaning toward her, comforting her.

She had never experienced such a feeling before and has not felt it since.

Did she hear it through the grapevine or the mycelia beneath her body? The easy answer is that she imagined the message. If so, no loss, because imagination, like dreaming, helps us release pain, expel it, and hold it up to the light. A connection to nature can heal us, especially in these times of profound grief.

LIKE CARRE, I'M WORKING to get to know the place where I live. This idea brings us back to the human senses, including the neglected ones.

To know a place is to smell a place. And to touch it, too.

As I mentioned earlier, as a boy, like most children who have a chance to be touched by the woods, I would lean forward to look between the deep cracks of bark. Doing that, I would smell more than I saw. Moss, fungi, or perhaps the tracks of ants. Wondering about that recently, I walked outside to smell a tree. I leaned close to the bark. Nothing. I was surprised. I leaned back and saw fire-produced scar tissue wrapped around the trunk. Maybe that was the reason I couldn't smell the tree. I moved to the next tree, which was unmarked by fire. Again, nothing.

This was an olfactory puzzle. Perhaps, in December, the trees had pulled back into themselves. Next time I hike, I promised myself, I would pay more attention to scent. But I'll probably look around first to make sure no one is watching me sniff the trees.

Unable to hear or see, Helen Keller interpreted the world through touch and smell, the latter of which she called the "fallen angel" of the senses. It "does not hold the high position it deserves among its sisters," she wrote in 1908. "The faintest whiff from a meadow where the new-mown hay lies in the hot sun displaces the here and the now. I am back again in the old red barn." She described how her sense of smell told her a storm was coming hours before it could be seen or felt. "As the tempest departs, receding farther and farther, the odors fade, become fainter and fainter, and die away beyond the bar of space."[58]

Writing in *InnerVoices: Empowerment and Personal Development for Introverts*, Jude Colin argues that "engaging with nature through touch isn't just a poetic notion; it's grounded in neuroscience.[59] As an ophthalmic technician, he points to research showing that "physical contact with natural elements can activate the brain's orbitofrontal cortex, an area associated with emotional and reward processing. This means that touching a leaf or feeling the bark of a tree can light up parts of our brains that make us feel good. It's like nature's own brand of mood-enhancing therapy." He makes a convincing case for touching plants or allowing them to touch you. "Our neurons, in fact, do seem to throw a little party, responding to the touch of nature with signals of pleasure and relaxation."

Peter Wohlleben, a German forester and author of *The Heartbeat of*

Trees, describes a game involving trees and touch. One person is blind-folded, and another guides them through the trees.[60]

"The goal of the short walk is a randomly chosen tree that the person wearing the blindfold has to touch," he writes. "The mossy cushions on the spreading roots, the texture of the bark, tiny twigs, the diameter of the trunk — all of these come into play." After the trek, the guide spins the blindfolded person, then takes off the blindfold. "The exciting part comes next," Wohlleben adds. "Will the person be able to find the tree?" Success usually follows. "One thing quickly becomes clear: our hands translate what they feel into pictures."

He reports that a 2014 study conducted by a team of international scientists found a direct connection between touch and vision. "Whenever the test subjects touched something with their fingers, their eyes stopped moving for a fraction of a second," he writes. "We are not aware of these minuscule pauses, but clearly, they are long enough to allow the brain to focus better and process what has been touched."

As we often grip, poke, and caress our cell phones, are we losing touch with touch? It's too early to tell. But just in case, go out and touch the earth. Or a tree.

Touch may not get the mainstream respect it deserves. But it does inspire efforts to reconnect with the earth.

A new phrase has entered the language: *Touch grass* is Generation Z slang used (seriously or snarkily) to tell someone to stop doomscrolling, step away from the screen, and go outside. A software developer has (of course!) created a Touch Grass app to help people get off their apps. To unlock access to an addictive social media site, you're required to touch some grass and upload a photo of yourself doing it. A reporter for *Outside* magazine used the app's calculator to estimate that he was "likely to spend 2,007 hours, or approximately 83 days, online this year. That's nine years over a lifetime."[61]

Then there's *grounding,* or *earthing,* defined by the website WebMD (which the online publication *Vox* calls the "most popular source of health information in the US" and "a hypochondriac's nightmare")[62] as a way to connect "to the natural electric charge of the earth, transferring electrons into your body."[63] Proponents suggest walking barefoot, lying on the ground, submerging yourself in water, and using grounding equipment such as grounding mats, socks, bands, and patches.

On Amazon, I counted seventy-two grounding products, including scores of shoes and socks. Some of these products come with an attached wire.

For maximum benefit, WebMD recommends that you plug the grounding mat or blanket "into a wall outlet with a ground or place the attached wire directly into the soil…If you plug one of these products into an outlet, make sure it's properly grounded."

Reading this, I felt my credulity meter twitch.

Grounding does have a loyal following, but studies are few and appear to lack scientific rigor. Still, over 1,500 studies suggest that contact with the natural world protects or improves health, no wires required.

IN THIS PRECARIOUS CENTURY, a new view of the natural world is emerging, one more inspiring than the mechanistic, reductionist definition of nature as a commodity or as a lesser heaven.

Elements of this vision come from traditional disciplines of science. Other parts are contributed by Indigenous people, traditional ecological wisdom, and the revival of ancient methods to connect with the natural world. Still more are provided by ecopsychology and ecotherapy, environmental education, and the growing nature-connection community.

At their extremes, each perspective is represented by passionate voices convinced that their approach is the only correct one. This attitude emerges naturally during times of anxiety and intellectual and spiritual upheaval. Still, by listening carefully, you can detect the shared language of longing.

Lyall Watson, a South African botanist and zoologist, attempted to explain much of what humans credit to the supernatural as, in fact, natural phenomena produced by what he called supernature.[64] That would be the mysterious *something* we sense in the forest but cannot fully detect or explain. So, we call it an elf.

Associating nature with the supernatural is tempting, but the term *supernatural* falls short. (*Supernatural* originated from the Medieval Latin *supernaturalis,* meaning "above nature.")

Can we imagine nature's spirit as an ethereal cloud above the mountain lions and the dandelions or lurking at the center of the earth? Of course.

We should also give credit where credit is due, not to the supernatural but to the mind-bending complexity of nature itself. Still, as illuminating as science has been, it remains limited in its ability to describe the fullness of nature. Art, however, extends science; it attempts to capture the inexplicable.

By waiting for all the evidence to come in, adults can miss out on the natural wonder that arrives before the hypothesis. Children remind us of that wonder.

The other day, while walking near our house, I encountered a four-year-old girl and her father. The little girl was pointing into the forest.

There, at the base of a black oak, was a delicate nine-inch door shaped with a pointed arch, just the right size for a fairy or a gnome to pass through. No one on the hill knows who is making and placing these enchanted doorways. More of them appear each week.

The little girl was beside herself with excitement and joy. Her unintelligible words were flying into the air. I caught her spirit. It was wonderful.

III. THE ART AND MUSIC OF NATURE

The real voyage of discovery consists not in seeking new landscapes but in having new eyes.

— MARCEL PROUST

CHAPTER 10

How to See a Flower

O NE MORNING, SITTING at a picnic table in front of Don's Market in nearby Santa Ysabel, I met with a young Ipai Kumeyaay woman with a striking smile. She told me about her art and her return to the land of her family and ancestors.

Now thirty-three, she was born on the Mesa Grande Reservation but raised in urban San Diego. When she entered the public school system and experienced racism for the first time, she fought depression, felt drawn to the reservation, and eventually moved there. After becoming a video artist, she adopted the name Fox Maxy. She later spent ten years in New York City, part of it working in the fashion industry, and in 2020, *Teen Vogue* magazine featured her as a rising star in the worlds of independent film and short video art.

Maxy now lives on the reservation with Peltier, her "badass dog."

She traded a blanket for Peltier at the 2016 protest against a planned pipeline near the Standing Rock Sioux Reservation in the Dakotas.

Recently, while studying the language of the Ipai, she learned that the Ipai word for *pet* also means "power," or "my power." She loves this idea because it is not hierarchical; it does not put pets below people.

Instagram and TikTok served as the venues for her first art videos. Over time, she built a collection of short, often abstract films and began to assemble them. Her skill and reputation have grown. In 2020, the COUSIN collective, which supports Indigenous film arts, awarded Maxy her first grant as a Native artist working in film. In 2022, she was named the Sundance Institute's Merata Mita fellow. Other grants and awards followed. Her films have been shown at the Museum of Modern Art in New York, the Toronto International Film Festival, the Los Angeles County Museum of Art, and the Lincoln Center for the Performing Arts in New York.

Her first feature-length film, *Gush,* premiered at the Sundance Film Festival in 2023. *Chanel* magazine described it as a "collage of audio-visual material — a kaleidoscopic study of horror and survival, an ode to friendship and a celebration of endurance."[65] She gathered the images for ten years; some depict the hills and life on the Mesa Grande Reservation.

Her work — and the way she talks — reveal a strikingly layered sense of place and identity, and demonstrate how she has absorbed the land and her ancestors. She tells me that some of her older relatives would walk fifty miles from the reservation to the coastal town of Oceanside.

"That's very old school," she adds, marveling at their strength. "It's nice to think about that when I'm going on my walks."

Now she speaks again of how this land and her ancestors drew her so powerfully. "All these memories, like my great-grandma's. I'm here because when she was little, she hid from Mexican soldiers trying to kill her."

Maxy also works part-time as a cultural monitor, a role created by a California law to protect Indigenous artifacts. Collaborating with paleontologists and archaeologists, she "watches the dirt" on construction sites, either before or during the digging.

She searches for shards of pottery, sculptures, stone tools, or "bones sticking out." She has learned to examine the ground surface for clues about what lies beneath. Certain plants grow where people have spent time: oaks and elderberry bushes, for example, because they were sources of food. At a village site, "people might have been working in one place for a long time. You might see rock chips where they were making tools or arrowheads."

So, for Fox Maxy, art comes to her from the dirt below her feet and the people she loves and her ancestors. She is a joyful person but carries grief for her people.

In a 2025 article in the *Los Angeles Times,* she writes, "People don't always know how to comfort each other. I think people have good intentions, but there's not a lot of education on how we can support each other in times of grief. People always say, 'Time heals everything.' But I don't know if grief ever really goes. It can be transformed into different types of feelings or energies, but I don't think it can ever just end or stop."[66]

She mentions the current mood of the country and captures fragments of that loss from the oaks, elderberry bushes, and her namesake, the fox, running toward shelter or freedom.

She turns all of that into her art.

WHEN I THINK of my First Place, I often see my father wearing his old army jacket and walking in the snow, with me trailing behind. Now and then, he would teach me something about the life cycle of, say, rabbits.

I also think of my mother at her art table. She was a greeting card artist. In my earliest years, I spent hours under that oak table. I read my books there. *Huckleberry Finn* or *Lad: A Dog,* or the pocket-sized *Golden Book of Reptiles and Amphibians.* It was a secure cave. Above the roof of that cave, my mother would hum as she drew and painted manger scenes with Baby Jesus, snowy hillsides, and reindeer floating across a dark sky.

Her life was challenging. When she was nine and living in Detroit, her father passed away. She and her mother moved from Detroit to the Mifflin family home in Independence, Missouri. Her mother worked as a practical nurse and rented out rooms. This was during the Great Depression. Times were hard. But my mother was a talented artist.

When she was sixteen, she went to work for J. C. Hall, founder of Hallmark Cards, in nearby Kansas City. She referred to Hall as Old Man Hall.

Occasionally, she joined the Hall family for dinner. On one of those occasions, the painter Thomas Hart Benton passed her the mashed potatoes. A master painter in the American Regionalism movement (and a self-declared enemy of modernism), Benton had helped shape the Midwest's view of itself and nature.

Benton, according to my mother, was a grumpy man.

In the late 1940s, with money she had earned at Hallmark and a $500 contribution from her uncle, she was able to enroll in the Pratt Institute, a school of art and design in Brooklyn. There, she met my father, who was studying drafting. He became a chemist and, later, a chemical engineer and helped build petroleum refineries.

She taught me that artists find patterns in nature. Sensory patterns and story patterns of what may or may not be there. The artist and the scientist may see similar patterns, but they observe, measure, and interpret them with different tools and ways of thinking.

Now and then, my mother would take a break from greeting cards and set up a chair and an easel at the edge of the yard where the woods began, and she would paint a tree in its season. In these paintings, you can see air moving and leaves lifting. She always left the work unfinished. She seemed happiest when she painted like that. Then she would return to the art table.

That table is now in my home office. Its surface is still (and will stay) covered with solidified globs of acrylic paint, residue from Scotch tape, and gouges from pushpins or blades. Soon, the pages of this book will be spread out on that surface for editing.

As a boy sitting under that table, I absorbed some ability, maybe through the wood. But my mother wanted me to look beyond the pinched cartoons I liked to draw, to the vast expanses of the earth and the big-brush world. One summer, she enrolled me in a painting class at Kansas City's Nelson Art Gallery (today the Nelson-Atkins Museum of Art). The class didn't stick. Stubbornly, I preferred Walt Kelly's comic strip, *Pogo;* the swirling bark, the kneelike roots, the currents of streams, the patterns — large and small — of his imagined Okefenokee Swamp. Nature was the most prominent character in his drawings. Now, in these mountains,

I glimpse the competing patterns that Kelly or my mother or my father might have detected.

Through my parents, I learned that art and science don't always get along, but they are not enemies. Rather, each is like a lens in a pair of binoculars. Each lens observes the same world from a different angle — and a lens can break.

In recent years, many scientific research institutions have shifted their focus away from observational learning and field studies. Decreased funding and changing priorities now favor computer science, biotechnology, and artificial intelligence over traditional fields like biology, zoology, and botany. The idea of observing the natural world doesn't seem as romantic as it did in the early years, when Jane Goodall used only a notebook, binoculars, and empathy to understand chimpanzees' lives. This move away from field observation impacts primary and secondary education and encourages the next generation to avoid hands-on experiences in nature.

However, the mechanical, reductionist view of nature might become an anomaly in Western society. Signs indicate that we many of us are already shifting back toward a more attentive, experiential approach to studying and experiencing the natural world.

In the late 1500s, philosopher and scientist Francesco Patrizi coined the term *panpsychism*, which comes from the Greek words *pan* ("all") and *psyche* ("mind" or "soul"). Panpsychism is the theory that consciousness exists everywhere and permeates the universe as much as matter, gravity, or electric charge.

Increasingly, we hear the ideas of panpsychism expressed by philosophers, religious leaders, artists, and scientists. As climate change increases, biodiversity declines, and eco-anxiety grows, more of us employ a range of nature-noticing methods — from the ancient art of foraging for wild food to reading the weather as a survival skill — often utilizing new technology.

IN HIS BOOK *The Urban Naturalist*, Dutch evolutionary biologist Menno Schilthuizen proclaims a rebirth of natural history, with new studies teaming up biologists, botanists, and other scientists with citizen naturalists.

This partnership brings natural history full circle: "In the nineteenth century," Schilthuizen explains, "scientists such as Charles Darwin and the men and women in his network were priests, notaries, doctors…None of them was a professional scientist. They did their groundbreaking research at home and in their nurseries and gardens. Fieldwork was done during family outings, and papers were written by gaslight at night."[67]

Traditional ways to study and connect with nature are regaining popularity; among them are gardening, farming, hiking, rock and mountain climbing, birding, and mushrooming (or "shrooming"). Hunting and fishing are holding on, although licenses for these activities are not selling as well as they did in previous years. Schilthuizen also points to the apparent rediscovery of the old-fashioned field notebook, the butterfly net, and the hand lens — as well as methods new to the scene, including smartphone apps that recognize birdcalls, insects, animal tracks, plants, and stones.

Dozens of new portals are opening, from soundscaping — mapping the sounds in, say, your neighborhood — to learning nautical navigation by the stars. "Imagine taking your smartphone-turned-microscope to an empty lot and discovering a rare mason bee that builds its nest in empty snail shells," he writes. "Or a miniature spider that hunts ants and carries their corpses around."

New forms of outdoor sports and games are also emerging. These include digital treasure hunting for objects hidden in forests and orienteering, a competitive compass-navigation activity.

WHILE CURRENT MEDIA COVERAGE and many educators emphasize the advances and importance of technology, art is still a doorway to natural history.

Centuries ago, the roles of artists and scientists often overlapped. In the seventeenth century and early eighteenth century, Maria Sibylla Merian painted the metamorphosis of caterpillars into butterflies. In so doing, she undercut the accepted scientific view that insects generated spontaneously from the mud.

Samuel Morse, who painted large neoclassical paintings, also invented the dots and dashes of Morse code. (In 1844, in his inaugural message sent from Washington, DC, to Baltimore, Morse famously tapped the biblical question "What hath God wrought?")

Leonardo da Vinci not only painted the *Mona Lisa* but also pursued botany and geology. Through his ageless notebook drawings, he foresaw the helicopter, the parachute, and the calculator. And he offered his generation and ours this advice: "To develop a complete mind: Study the science of art; Study the art of science. Learn how to see. Realize that everything connects to everything else."[68]

The Nobel Prize–winning physicist Richard Feynman often spoke of an artist friend with whom he disagreed. "He'll hold up a flower and say, 'Look how beautiful it is,' and I'll agree. Then he says, 'I as an artist can see how beautiful this is, but you as a scientist take this all apart and it becomes a dull thing.'" Feynman found that view "kind of nutty." He believed that he could see much more about the flower than his friend could. He continued:

> I could imagine the cells in there, the complicated actions inside, which also have a beauty. I mean, it's not just beauty at this dimension, at one centimeter; there's also beauty at smaller dimensions, the inner structure, also the processes. The fact that the colors in the flower evolved in order to attract insects to pollinate it is interesting; it means that insects can see the color. It adds a question: Does this aesthetic sense also exist in the lower forms? Why is it aesthetic? All kinds of interesting questions which the science knowledge only adds to the excitement, the mystery, and the awe of a flower. It only adds. I don't understand how it subtracts.[69]

Great artists and great scientists look beyond the studied thing. In her own ode to a flower, Georgia O'Keeffe described perception as both shared and individualized:[70]

You put out your hand to touch the flower—lean forward to smell it—maybe touch it with your lips almost without thinking—or give it to someone to please them. Still—in a way—nobody sees a flower—really—it is so small—we haven't time—and to see takes time, like to have a friend takes time. If I could paint the flower exactly as I see it no one would see what I see because I would paint it small like the flower is small.

So, I said to myself—I'll paint what I see—what the flower is to me, but I'll paint it big, and they will be surprised into taking time to look at it—I will make even busy New Yorkers take time to see what I see of flowers.

Well—I made you take time to look at what I saw and when you took time to really notice my flower, you hung all your own associations with flowers on my flower and you write about my flower as if I think and see what you think and see of the flower—and I don't.

MY MOTHER AND FOX MAXY came from entirely different backgrounds. But I think they would have been able to talk for hours about art and place. The same is true of artist and illustrator Lucy T. Smith. Today, she helps botanists at London's Royal Botanic Gardens identify new species. On a podcast associated with the journal *Nature,* she reports, "We also feed back to the scientists and say, 'I've seen what you've asked me to see. But do you know what, I've also seen this. Did you know that this flower has this structure?'"[71]

Clearly, there's more than one way to see a flower.

And more than one way to save the natural world.

Until he passed away two years ago, my friend James Hubbell lived on a hill just west of Volcan Mountain. As an artist, he inspired countless people to find beauty in nature. In an interview, he described being at London's Tate Museum in 1982, during the Falklands War between the United Kingdom and Argentina, a conflict over territorial control of the islands: "*Everybody* was talking about it, even the guards."[72]

At that moment, he was looking at the work of John Constable, a nineteenth-century English landscape artist. "And I remember standing in front of this little drawing by Constable of a wave," he said. "*It was so beautiful.* I thought, why in the world would anybody be interested in a war when they could look at a line like that! I think — and I believe this — that it's possible to draw a line that changes the world. In a sense, it's that simple, but how do you draw that line?"

"I Notice, I Wonder, It Reminds Me Of"

Hiking with Marley Peifer is like taking a walk with a three-year-old. In a good way.

Every few feet, he stops in his tracks and points excitedly. "Wooooo! A cool coleopteran! A *cooperative* coleopteran!" By that, he means the beetle is not moving, so he can drop to his knees to take a closer look.

"Native bees!" he says. "They're all around us. See how they feed on that flower? That bee is holding on to the petal with just its mandibles. Some bees do that. They sleep that way. I've seen that in photos, but this is the first time I've seen it in real life."

We're making our way up the trail on Volcan Mountain.

Marley Peifer is an up-and-coming YouTube evangelist for the fast-growing practice of nature journaling — that is, the practice of taking a blank-page journal, along with pencils, pens, brushes, and cameras, into a natural habitat and recording life through paintings, drawings, prose, poetry, and photographs. Some journalers even include links to videos, field recordings, and soundscapes gathered in the field and published on their website.

The location can be a mountain, a city park, or the far end of a suburban yard. Nature journaling is inexpensive, helps diffuse biological knowledge, and promotes contemplation. It can be pursued with a group or in solitude, at any age. It's a perfect tool for noticing.

Marley and I have, by now, passed through an arching, sinuous sculpture by James Hubbell. The piece is partly made of river stones and chips of glass melted by the most recent major fire. Marley heads up the trail again but goes only a few feet, because he encounters a wild cucumber — aka *Echinocystis lobata,* or spiky cucumber. "Look," he tells me, "I'm easily distracted when I notice shiny things."

He kneels and touches it.

Marley is in his thirties. He moves quickly. He is wearing fingerless gloves and a straw sombrero. A loupe, a small magnification lens, hangs on a cord around his neck; his art supplies and nature journal are wedged into a pouch slung over his shoulder.

Later, he takes the journal out. He flips through it. Most of the pages are still blank, but the early pages are filled with sketches, finished drawings, watercolors, and handwritten text. Bugs, birds, plants, seeds. In large, ballooned, multicolor letters, he has drawn the word *Opuntia* (commonly known as prickly pear cactus) on one page. Covering most of another page, he has inked in a leaf-footed bug, with notes: "Has a stink gland!"

Growing up in San Diego, Marley often hiked and camped in these mountains. He feels they're part of him. They are part of his First Place.

He lived next to the city's Tecolote Canyon, one of the many watershed canyons that lace through the city and county. Marley immersed himself in the canyon and his father's vegetable garden. Then, when he was seven, his parents divorced, and his father moved to Costa Rica. Marley would spend part of the year there and part in San Diego.

Marley's bond with the natural world emerged mainly in Costa Rica. "Everything was hands-on. We played hunter-gatherer games. We'd fish for crawdads, pick fruit, and chop things up with our machetes; that was how I started getting into nature connection. Not guided by adults, for the most part," he says as a rising wind kicks up puffs of dust on the trail ahead of us. A red-tailed hawk circles.[v]

"I love this lupine," Marley says. Then he adds, "I notice, I wonder, it reminds me of." That's the core of his teaching and his mantra.

A COUPLE OF YEARS EARLIER, I spoke with author and educator John Muir Laws, the most famous nature journaling educator and proponent. Laws (unrelated to the naturalist John Muir, who founded the Sierra Club) is partly responsible for the quick rise of the avocation. In 2022, he cofounded the Wild Wonder Foundation to help teachers and students learn about the benefits of nature journaling.[73]

In 2016, he introduced author Amy Tan to the practice. That year, as she writes, "racism against Asians became more blatant."[74] She found refuge in nature and in the art she had loved as a child. Later, she assembled her journals filled with drawings of birds into her bestselling book *The Backyard Bird Chronicles*.[75] David Allen Sibley, ornithologist and author and illustrator of *The Sibley Guide to Bird Life*, wrote an admiring foreword: "The birds that inhabit Amy Tan's backyard seem a lot like the characters in her novels."

"To draw portraits of the birds," Tan writes, "I had to feel the life within and see what the bird was seeing, feeling, and planning as its next move. I imagined what it thought about me. I thought about survival."

Like Marley Peifer, Laws emphasizes slowing down. In the field, he advises his students and acolytes to "keep engaged with the same sunflower, push through the veil where we usually stop paying attention." An eloquent emissary and a mentor to educators like Marley, Laws considers nature journaling to be "reverent attention...the single most powerful tool to supercharge your observation, memory, and connection with nature. It is the critical foundational habit of being a naturalist and scientist."[76]

---

v   Chances are, if you hear the call of a bald eagle in a movie, the sound is actually that of a red-tailed hawk. A bald eagle's cry is not nearly as appealing.

Paying attention, says Laws, "is what makes you fall in love with the world. It is through attention that we create memories, but those memories will change over time."[77]

We still pay a price for Descartian certainty that animals and plants are enlivened machines. Or that only humans have a soul or know love or loss. The reductionist view of nature discourages certain types of scientific speculation and the imagination of the young.

Education budgets are tight. Tempers are short, and no one wants to be considered superstitious or too far off the sidewalk. Consciousness in nature? That's for animal whisperers, clairvoyants, and tree huggers who *talk to trees.* In their book *Wonders and the Order of Nature, 1150–1750,* science historians Lorraine Daston and Katharine Park write, "Wonders, as well as marvels and miracles, were central to the medieval and early modern world views, but since the late eighteenth century they have been subordinated to Enlightenment reason: 'To be a member of a modern elite is to regard wonder and wonders with studied indifference; enlightenment is still defined in part as the anti-marvelous.'"[78]

Art can remind us that nature is worth more than what the market will bear. Nature journaling reminds people to use their senses, then commit what they have noticed to paper.

MARLEY HAS MOVED farther up the trail. I mention what I learned earlier about fairy rings. As he kneels to feel the texture of dry grass, he cautions that negative interpretations of fairy rings — that the fairies are luring us into danger — may be an overlay by Roman and, later, Christian colonizers to translate the stories of those earlier cultures into the devil's work.

From the horizon, clouds spread like reaching fingers moving slowly above us.

Marley is limping slightly today from an injured knee. Up the path, he stops and drops to his knees again, ignoring the pain.

"Wow! Look," he says, pointing to some small yellow plants. "It looks like a parasitic plant, like it's coming out of an anthill." He takes a photo for iNaturalist, one of several apps that collect images from around the world and match them with requests from hikers and others attempting to identify plants or animals.

We walk on. "I never connected drawing or art to the natural world," Marley says. "It wasn't until after college that I learned about a scientific illustration graphics program and applied for it. That's when I discovered John Muir Laws. I didn't get into that specific program, but I became hooked on nature journaling in 2014. I realized that nature journaling accelerated my learning in other areas."

In no time, we encounter another Peifer stopper.

Once again, he is on his knees, pressing a twig into a fist-sized pile of animal scat. "The first thing I always want to do is push on it and see how hard it is," he says. The poop is new, he decides, because it is soft. "It's also pretty homogeneous and dark. If it came from a coyote, the coyote isn't getting much in its diet except animals."

At the sight of this homogeneous poop, I remember taking a short hike with a park warden at Prince Albert National Park, in the Canadian province of Saskatchewan. Like Marley, the ranger was a stopper. And a pointer. "Wolf scat," the warden said. The scat was large, dry, and chalky white — an indication its creator was a carnivore who wouldn't eat their vegetables. I pulled out my phone, took photos of the Canadian poop, and immediately sent them to Kathy and friends, sure they would be as thrilled as I was.

Marley, still kneeling, pushes on the poop again. "There's bone right there," he says. "Looks like a tooth. Look at that. You see that? Looks like a carnassial, too."

Carnassials are the enlarged upper and lower teeth of a carnivore, shaped and self-sharpened like the blades of scissors. These teeth pass each other in a shearing action. Marley surmises this scat was left by an animal that ate flesh. Marley also thinks he sees a shell. Or a scapula. He pauses, then says, "No, it's a skull."

We climb a couple of switchbacks, and again he stops — suddenly, of course — and points to an ant crossing the trail. "I notice an ant," he says, "but nature is weird; it might not be an ant. Sometimes an 'ant' can be an ant mimic. Several other arachnids and insects imitate ants. Sometimes flower petals aren't flower petals but plants with floral parts that imitate insects. Which helps attract pollinators."

I look closer at the 'ant.' "Is that a honeybee?" I ask.

He squints. "It looks more like some kind of small butterfly." His advice: "When you see something interesting, you say to yourself and maybe to the people with you, 'I notice something red that looks like an insect crossing the trail. I notice it has six legs. Let me double-check.' You focus on direct observations, not interpretations. An interpretation is the result of all these small bits of visual information that your senses are picking up. And your brain is jumping to conclusions." Our ancestors probably found jumping to conclusions useful, he adds, "because they didn't have time to get all academic." Getting that Latin name right isn't important when a cave bear is chasing you.

While looking for signs or tracks (which he used to do), he finds it too easy to get caught up in confirmation bias. "Once you come up with an idea, your brain filters out information that contradicts your idea," he explains. "Another thing that happens is when you're with a group of people, the person who is the leader or the adult who has memorized the most factoids just spouts out information — and everyone turns into a consumer of that information." They quit noticing.

Everyone can make a valid observation, he adds. Distinguishing between observation and interpretation can be challenging, but it becomes easier with experience and self-awareness. For example, Marley compensates for his preference for eagles and hawks by consciously focusing his attention on other birds, to hear and see them more clearly.

Assumption of knowledge, especially superior knowledge, is another barrier to noticing. This is true for both adults and children. At times, it's better for a parent or a teacher not to know, or not to pretend they know, the answer. Then the child and the adult can ask questions together and share the awe of discovery. Later, the child is likely to recall that moment of wonder more than an adult's quick, technical answer.

In Marley's world, asking questions is valuable not only because they retrieve more accurate information but also because they nurture humility. "What I try to do is to teach metacognitive tools — thinking about thinking. If I can help people slow down, we can daylight some of the unconscious processes that can help them notice more." But before he teaches "conscious noticing," as he calls it, he emphasizes "unconscious noticing."

"For example, out of the corner of my eye or your eye, an animal comes into sight. Your brain hasn't realized yet the importance of that animal. You've picked up olfactory cues that neither of us has tuned into so far today, things we feel on our skin, or you might have a bad feeling about this trail. You say, 'I have a feeling, let's go back.' And later, you find out that a wildfire had started just a few minutes earlier. It was beyond what you could see. The smoke had not yet reached your consciousness. These are the starting points, before I say, 'I notice.'"

Marley also confirms what I have been learning about direct and indirect vision. When looking directly at something, you find it easy to rationalize what you're seeing, because you assume you know exactly what you're looking at — and you don't.

Cue the humility.

Marley moves to the side of the trail and points toward the soft, sandy soil. "Looks like a lion track, right? Now I move over here and look at it directly, then move my vision away until it's in my peripheral vision, and then return to my direct, focused vision. Several times."

It is probably the track of a large dog.

"Even what we're doing right now, using words, is externalizing our thinking, which is good, because it allows us to ask questions instead of leaping to conclusions." Drawing the track in a journal makes it easier "to take our thinking outside of our brain."

On the southeast side of the trail, from nearly a mile high, we look down the steep, wooded slope at the outer buildings of Julian. Turning to the south, we see the distant mountain ridges of Mexico. Turning to the east, we see Mount Woodson and Iron Mountain, layered in gray and blue, and beyond that, lines of white clouds rolling in like ocean breakers lit by the sun.

*   *   *

BY MIDDAY, MARLEY has introduced me to a dozen species of plants, including the strange, spiky wild cucumbers and a stretch of white sage. "Indigenous people burned sage in spiritual rituals," he says. "Now you can buy it at Walmart and use the smoke to bless your house. As a result, sage is disappearing from the mountains."

Then he points to a spider's funnel, some mating assassin bugs, and a towhee with its mouth open wide, practicing avian air-conditioning.

I ask him if he sees faces in the bark of trees. "Yeah," he says, "but mostly I look for the bark and what it tells me. Was it trimmed or pruned, and where? And the wind information that the bark carries, how it leans and twists and breaks."

Marley reads his students like bark. Two kinds of students give him pause: the ones who believe they know it all and those who think they cannot draw.

The students who lack artistic confidence often become his best pupils. Beyond their usual comfort zone, they become more at ease with the unknown. Their pencil lines begin to appear almost spontaneously, arriving as poems, prose, or scientific notes. Over time, the notebooks themselves become works of art.

A few hundred feet below Volcan's summit, Marley is distracted by high grass on his left. He reaches out and pulls a stalk toward his eyes and studies its head, which is bursting with long spines.

"See that?" he says, pointing to a cluster of what looks like dead ants at the base of each cluster. "Those are tiny mushrooms called ergot."

Ergot grows naturally on grains such as rye and contains several poisonous and psychoactive alkaloids, including a chemical precursor to LSD. When consumed, the fungus can cause ergot poisoning, leading to symptoms such as headaches, vomiting, diarrhea, and even gangrene of the fingers and toes. Fortunately, flour makers remove ergot from grain.

Marley tells me that some historians speculate that an ergot infestation in 1691 Massachusetts may have contributed to the Salem witch hunts.

He pauses. "Cattle get into that grass, they have a bad day."

CHAPTER 12

⫷ Susan and the Dragonfly

THIS AFTERNOON, KATHY and I walked up the road toward Pine Hill Lodge. The lodge is a shambling structure built in 1912, when woodsy Adirondack architecture and Teddy Roosevelt were in vogue. Next to the lodge is a weathered, barnlike building said to have been built as a training gym for the boxer Jack Dempsey in preparation for his 1926 match with Gene Tunney. Records are vague about whether Dempsey used the Pine Hill gym, but confirmation bias says he did.

Late-afternoon light slanted under the pine and oak branches. Native toyon bushes, along with the fingers of red-barked manzanita and white sage, filtered the light. (Okay, I looked those names up. I'm still a flatlander.) We were talking about what we see.

"When we walk, I usually don't see anything new," Kathy said, "but I always see things that are different. My eyes are attracted more to plants. Yours are drawn to animals."

True. Occasionally, I leave my glasses behind. A few weeks ago, I asked our optometrist if living in a place of distant horizons improves vision. "Very possible," he said.

My peripheral vision also seems to have become more acute. Or that's what I like to think. I seem to see smaller movements in the branches. I look for eyes.

Spending the day with Marley Peifer may have also helped my vision or, more specifically, sharpened my perception of vision.

Artist Susan McDonnell did the same, but from the perspective of a professional painter and fine artist. She sent me an email and attached a painting of a dragonfly. She described how, in the early 2000s, she lived in a Minnesota home with a small garden pond.

"In the late Spring, a red darner dragonfly took up residency," she wrote. "I started taking photos at a distance…Over time the dragonfly let me get within inches and then let me lightly touch its wings. This dragonfly showed up every morning around 10 and patrolled the pond until around 4 pm for about 4 months. We 'visited' every day, and I spent a lot of time quietly observing the dragonfly, marveling at the dragonfly's beauty."[79]

The dragonfly, she added, "was a bit of a ham."

McDonnell was already an accomplished artist, but she considered her moments with the dragonfly to be transformative. "It was at this time my paintings became more focused and detail-oriented in subject and technique," she wrote. "All that slowing down, observing and being completely delighted every time I went to the pond and saw the dragonfly had returned had a huge influence on my paintings from then on."

The natural art of seeing is about resilience and transcendence, about not being trapped in time or your own body. For my prior books, I've interviewed artists and scientists who have learned how to look deeply at life, and through that looking, they have found their resilience.

At the Scripps Institution of Oceanography, Paul Dayton told me how he teaches students about the "X-ray" tradition of Australian Aboriginal art, which exhibits an elaborate understanding of the bone and organ systems of animals, "even showing the split lengths of snake lungs." He marvels at their mythology of animals, plants, and even rocks, as well as their concept of dreamtime, where all life and stories exist forever. "These people were and are spectacularly accomplished naturalists. When you look at those images, if you let yourself fantasize, what do you think these animals dream about?"

He challenges his students to go beyond the lab, to have the courage to dream about the lives of other animals, to use the abilities they were born with but may have misplaced in a society and education system disconnected from nature.

We can apply these concepts in our own lives. Here is a moment for our children, for you, to *look*—to paint, draw, or photograph what you see. Keep an indoor/outdoor illustrated journal. Set up a table next to a window, with art supplies. In her book *The Nature Connection,* naturalist and artist Clare Walker Leslie writes, "I often draw little landscapes in places where nature seems hard to find—out a classroom window, along a highway, or outside a hotel. Do you remember 'Where's Waldo?' If you look, you'll be surprised at what you might find."[80]

Beyond seeing is becoming. The nineteenth-century Postimpressionist painter Paul Cézanne described his version of becoming: "The landscape becomes reflective, human, and thinks itself through me. I make it an object; let it project itself and endure within my painting…I become the subjective consciousness of the landscape, and my painting becomes its objective consciousness."[81]

A FEW MONTHS after sharing her dragonfly painting, McDonnell sent me a series of reproductions of paintings that she hopes to use in a children's book. Like Bateman's bears, her bears, rabbits, and foxes seem to quiver with life; their eyes suggest thoughts or what exists between their thoughts.

Curious to know more about her process, I called her. She had moved to Northern California and continues to pursue animal encounters for her art.

Describing her approach, McDonnell says she goes into the woods or some other natural habitat, selects a spot to sit and watch, and listens, feels, and imagines. She describes her modified sit-spot method as "going to a kind of childlike place, a world where you look and marvel at what's unfolding in front of you." She feels that she almost disappears. Then the wildlife shows up: often birds and bugs and then the larger mammals.

"Words aren't my favorite things," she says. "I've tried sometimes to express this, but I do feel as if I disappear. I mean it. It's an immediate engagement. It almost doesn't matter what's on my mind; I can get there fast. In my mind, I turn into a little kid, like Alice seeing through the looking-glass. I'm having a conversation with this other being. I'm not saying the animal is *talking* to me, but it does give me certain looks, like, 'Who are you? What are you still doing here?' Finally, the conversation goes away, and that's when time collapses." Typically, all of this occurs within a few seconds.

She brings her camera to "capture that look, that expression, that moment of connecting that I can bring back to the painting." Or more accurately, *forward* to the canvas. A gallery owner who represents McDonnell's work in Santa Barbara suggested another word, rather than conversation or connection: *communion*.

"I know that word is loaded, but I think it comes close," McDonnell says. "Communion. The word 'communion' comes from the same root as 'community.' A larger community." One that includes all species, not only ours.

McDonnell doesn't always feel so enlivened or spirited when she's in nature. Sometimes other feelings do intrude, not from the world beyond the path she's on or the animals she encounters but from the vulnerability of being human. "That negative feeling could point to the unseen dangers that are in the woods. Or maybe that feeling comes from my personal demons. I've turned my back to them, and they run and catch up with me. Which is one reason that, as best I can, I try to empty myself and not bring anything with me out there."

I mention Robert Bateman and how he becomes the bear. She feels something similar. "I'm not sure why, but the animals in my paintings seem lost in thought. They're pausing in the middle, from here to there." She says she doesn't pretend to know what they're thinking, and she doesn't want to disrupt their thinking. "But often we lock eyes, and the separation between the two of us disappears. I fully believe they have consciousness. They don't necessarily worry about the future." For her, these moments are similar to being with "a friend or relative who somehow makes you feel good about yourself, when you're with them."

This assumes, of course, that the wild animal means her no harm.

Next, McDonnell uses her imagination. Her description brings to mind etuaptmumk, the two-eyed seeing described by the Mi'kmaw elder Albert Marshall, or William Blake's description of twofold seeing.

As she watches the animal, she asks herself about the animal's story.

What did it have for breakfast that morning? Who is its mate? Has it been fearful or brave or both that day? "When you catch them off guard like that, they are in between things. They're having a thought; things are going through their mind, and those things show up in their eyes. That's when they show their consciousness."

For McDonnell, those moments are akin to being in the creative zone or a state of flow. She loses track of time. The effect happens twice, first when she's watching the animal, and once more when she's painting it. "I disappear again. I can feel this animal's presence once more."

So can the viewers of her art.

When I look at the eyes of the rabbits and birds she paints, I can feel their thoughts and their stories. The same is true for Bateman's bear when it steps out of the wall. Do the stories belong to the bear or Bateman? Does it matter?

Artists are licensed to imagine. Scientists are, too, though some may have trouble with that license. It's too loose in their pocket, too open to interpretation and projection. But scientists also see stories. Research in environmental neuroscience suggests that spending time in natural environments stimulates and sharpens both imagination and empathy.

For noticing and interpreting the natural world, the value of imagination, empathy, and even anthropomorphism is underrated. Each quality has its dangers and capacity to mislead. Each contains possibilities for wonder and the transformation of Earth and of us.

Several years ago, I enrolled in a program called A Sense of Place. The basic idea is to help people come to know where they live. Usually, a dozen or so people arrange for an expert—a tracker, a wildlife biologist, an archaeologist—to walk with them on a trail of their choice, one steeped in history or nature. The expert shares a walking lecture with questions and answers. It's a terrific format. Beats sitting in a lecture hall.

One of our lecturers was a geology professor from the University of California, San Diego (UCSD). He chose a path on the beach under the high sandstone cliffs of Del Mar, California. As we walked, he described the layers of sandstone, what the different colors of the rock revealed, and what fossils could be discovered there.

He stopped for a moment and said, "In all these details, I don't really see rock or sand anymore. I see a story. One that spans millions of years. Every geological formation is telling a story. Its own story. And our story."

The Photographer's Eye

W E CAN'T ALL become fine artists, but we can become the bear. Or the dragonfly just beyond the windowpane. Because of art, we see the world through other eyes.

For many years, I've had an affinity for photographers as people and as professionals. They're generally more relaxed than reporters and usually great companions on assignment. I envy them their immediate gratification, how they so often know in the instant of the shutter's release that they have captured — or perhaps freed — a moment in time.

For amateurs and professionals, photography can be one of the best ways to notice and then share nature. Photography is more familiar to most people and faster to learn than oil painting or even nature journaling. Digital photography has exploded in popularity.

"The best camera is the one you have with you," the saying goes.

For many of us, that means our cell phone — which offers an added advantage: no more film or darkroom chemicals. Just a memory card and a computer.

"To 'see' according to ecological psychology . . . stresses that patterns of light on the retina change relative to your movements," writes science journalist Amanda Gefter. "It's not the brain that sees, but the whole animate body. The result of 'seeing' is never a final image for an internal mind to contemplate in its secret lair, but an adaptive, ongoing engagement with the world."[82]

The photograph begins before the click.

Occasionally, photography is dismissed as one more technological barrier between a person and a genuine connection with nature: Put down the camera and look. There's some justification for that when the f-stop setting becomes more important than the fox stopping in front of you.

But others see the camera as an essential field tool. They argue that, in fact, the camera, when used appropriately, reveals what we might otherwise miss.

Taking the photo slows us down so we *can* look. In some settings, a camera can bring us closer to nature than painting or drawing. At least that is the conclusion of researchers at the UK's National Trust and University of Derby in their paper "Actively Noticing Nature (Not Just Time in Nature) Helps Promote Nature Connectedness."

The authors make a good case: "Photography not only exposes people to nature whilst capturing the image, but also when reviewing and editing the resultant files, with close attention being paid to the features portrayed in the image."[83]

Here's an example. One day, I was sitting at my computer, reviewing photos of rock patterns and tree bark. I'd enjoyed taking the photos in the field. But now, I was startled by what I had not seen when I took the picture. Hidden in the bark was an eye looking back at me. When I posted a photo on my blog and asked readers, "Just whose eye is this, if it is an eye?" the answers came: A bird. A lizard. An insect. Nothing at all—what eye?

One mother said she showed the photo to her fourteen-year-old son, who said the solution was obvious: The eye belonged to a *dragon*. Of course it does! I'm with the kid.

PHOTOGRAPHERS DESERVE SIGNIFICANT CREDIT for the creation of the national parks. Sometimes a photo is worth a thousand filibusters. National parks and other public wildlands are today threatened by politics. These

threats make conservation photography more critical than ever. The Wilderness Society makes the case: "Since the 1860s, photography has played an important but little-known part in the field of nature conservation." The society points to "a handful of pioneering artists [who] intentionally used their photographs to promote conservation."

Among the most famous are Carleton Watkins (1829–1916) and William Henry Jackson (1843–1942). According to the Wilderness Society, Watkins's photographs of California's Yosemite Valley influenced the decision by the US Congress to pass the Yosemite Grant. With this measure, Yosemite Valley and the Mariposa Grove of the Giant Sequoias became the first lands that Congress established for public use and preservation.[84]

No one has done more to attract people to nature photography and the landscapes of the United States than Ansel Adams (1902–1984). As one of the great noticers of any era, Adams focused on the big picture: the great western expanses, the night sky above a graveyard capturing the last light, the bare flanks of Yosemite's El Capitan. In his autobiography, Adams wrote, "There are certain times, when, as on the whisper of the wind, there comes a clear and quiet realization that there is indeed a presence in the world, a nonhuman entity that is not necessarily inhuman."[85]

Other types of photographers, including tourists with their Kodak Brownie cameras, continued that tradition. Then came the single-lens reflex and now its digital equivalent.

Today, most of us carry smartphones. Despite their capacity to pull us away from nature and each other, most phones on the market offer excellent built-in cameras. In part because of this capability, wildlife photography has expanded beyond *National Geographic* spreads to countless Instagram posts and Facebook nature photography groups. Intentionally or not, these amateur photographers have sparked increased interest in, and empathy for, the other wild places: the streams that run through suburban neighborhoods and the backcountry, the birds of a million backyards, and the landscapes that have escaped the notice of so many.

Amateurs also record the health of usually invisible wildlife alongside professionals who continue their work, despite economic competition from digital pixel inflation. Notably, the earliest amateurs were those who could afford expensive cameras and darkroom equipment.

Marine biologist, photographer, and author Cristina Goettsch Mittermeier coined the term *conservation photography*. She also founded the International League of Conservation Photographers in 2005 and continues to play a role in the appreciation and protection of natural habitats in the wild and in cities. Amateur photography has continued to influence conservation, just as it did the civil rights movement.

A few years ago, I met Melissa Groo online and quickly became a fan of her work. Since then, we've continued a conversation about the role of photography both in conservation and as a means of entry into the other worlds of wild animals.

Groo is among the country's best-known conservation photographers. Unlike Adams, she favors animals over landscapes. She also writes a column on wildlife photography for *Outdoor Photographer* magazine and serves as a contributing editor to *Audubon* magazine. At the Cornell Lab of Ornithology, Groo created the institution's first online bird photography course.

Groo usually calls herself a wildlife conservation photographer, but nearly as often, she refers to herself as a wildlife biographer, because her cameras make her notice and share animals' stories.

Born in 1962, Groo grew up in New York City on Fifth Avenue across from the Metropolitan Museum of Art. She lived in an urban environment and had limited contact with wildlife. But her great loves were her cats and dogs, which, she told an interviewer, taught her much about the individual personalities of animals. She loved *Watership Down,* by Richard Adams, which describes a fictional civilization of rabbits.

Through her photographs, Groo has told the stories of endangered giraffes in Uganda, spirit bears in British Columbia, sandhill cranes in Nebraska, snowy owls in Alaska, and the Hudsonian godwits of Chile. She has found herself remarkably close to the pumas of Patagonia and leopards in Botswana.

I asked her about her photographer's eye and her advice to amateurs.

Groo encourages all photographers — including smartphone users — to use their talents to promote the protection and expansion of natural habitats. Like artists Susan McDonald and Marley Peifer, Groo urges photographers to put themselves in the place of the animal, to think and feel how the animal might think or feel. "Steep yourself in the world of the animal," she advises. That also applies to trees, plants, insects, fungi, and other forms of nature.

Skilled nature photography is about more than the photographer's eyes.

"Oh my God," she says, "it's so much about sound. Weather sounds. The sounds your subject makes. The sounds that may have to do with alarm." By observing bird behavior and listening closely, she says, "you can learn that another bird of the same species is in the area, probably a mate." And listen to other bird species, or squirrels sounding their alarms.

Often, she hears before she sees. Crows are a tip-off system, she adds.

When Groo hears a crow ruckus, she gets out of her car, goes into the forest, finds the crows, and seeks the reason for the birds' broadcast. One recent example of the source of the racket was a great horned owl. The owl did not meet with the crows' approval. "The crows were going off like fireworks in the trees," she recalls. Groo lifted her three-foot-long telephoto lens and photographed the owl, then headed back to her car.

In 2000, Groo worked in Central Africa alongside Katie Payne, an elephant researcher from Cornell University. Payne pioneered the study of the acoustic communications of forest elephants. "We were deep in the rainforest studying them and collecting their sounds," Groo recalls. To do that, they used equipment that could detect infrasound — soundwaves with frequencies usually below 20 hertz, which humans cannot hear. Although infrasound animal communication is inaudible to humans, she says, "you can feel it, like when a car drives by with windows rolled up, stereo booming low notes. It's like a throbbing in the air."

"Elephants use infrasound because it can travel longer distances — through the forest and over the savanna," says Groo. "They communicate with other elephants, coordinating their movements, or sharing the sources of food, or staying in touch when family members are separated."

Over time, she learned to sit for hours — even while being attacked by stinging sweat bees — watching behavior unfold, often slowly. She also learned the importance of knowing the natural history of an animal, which helps the photographer *see and sense* what the animal might do next.

While she believes that wildlife photography can help protect and conserve species, her primary motivation is personal and internal. "When I'm out photographing, I feel more alive than at any other time in my life. It's the great escape, to be tuned into the natural world, to let everything else, including news and social media, fall away." She also believes that nature is the ultimate social medium. Stories are unfolding around us, even in cities. "A pigeon's cooing may seem mundane to some, but others love the iridescence of an urban pigeon's head and throat," she adds.

In addition to her other professional work, Groo serves as an adviser to the National Audubon Society on photography content and ethics. In her columns and presentations, she expresses anger at the growing trend toward "fauxtography," which includes staging wildlife photos, attracting animals, and modifying their behavior through bait (such as food, bird-call recordings, and other techniques). Staged photos harm animals by disrupting their natural behavioral patterns and undermining the authenticity of their experiences.

"Just by being out in nature, I disturb animals," she admits. "It is incumbent on me to arm myself with knowledge that will help to lessen that disturbance. That means knowing how the animals I set out to photograph nest or den, forage/hunt, feed, rest, and raise young."[86]

She also learns their vocalizations of alarm or stress.

"I study the ecosystem and learn what features of the landscape are useful to the animals that live there, for concealment, protection from the elements, and food." She learns about their predators — and, intriguingly, how predators learn about their prey. "Will my presence, or a sign of my presence after I leave, be an alert to those predators?"

She tries to blend into the environment, often using pop-up blinds or her car as a blind. If she does approach an animal, she first gathers information on the animal's baseline behavior from a distance before moving in closer. Sometimes, if the animal does not accept her in its territory, even temporarily, she leaves. She doesn't use audio lures for birds; she makes no noises to attract any wild animal. "I don't believe that any wild animal should ever have to perform for us," she adds. She never feeds wild animals, nor will she photograph animals being fed by other people.

Her overarching principle is simple: Keep wild animals wild.

"That's the most responsible and the kindest thing I can do for any wild animal," she says. "This is especially true when it comes to predatory animals like owls, wolves, foxes, and coyotes, for whom habituation to humans is often a death sentence."

Another popular form of nature photography is the use of autonomous trail cameras — or camera traps (which employ professional-grade cameras). Not yet a passing trend, trail cameras and camera traps open new worlds and build admiration for how wild animals survive.

Melissa Groo has placed a few camera traps near her home. She sees little evidence that they cause disruption so far, but she remains vigilant. When she collects trail-cam images and views them on her computer, she looks for photos that depict startled animal reactions or distress. For the most part, the animals ignore the cameras and continue their daily or nightly business. "Sometimes, I have seen raccoons that appear visibly startled by the flash," she says. "They pause for a moment or pick up their pace. But they don't run away."

Even so, she dials down the flash intensity of her trail cameras.

IN SOUTHERN CALIFORNIA, Ray Toft is one of the most accomplished camera trap photographers. He lives a half hour west of Julian in the dry mountains near the town of Ramona, a city named for *Ramona*, the popular 1884 novel by Helen Hunt Jackson. The book, about the mistreatment of Native Americans, is sometimes compared to Harriet Beecher Stowe's *Uncle Tom's Cabin*.

With a population of over twenty-two thousand, Ramona sprawls across scrubland and farmland. On the sides of Ramona's commercial buildings, murals depict a way of life from a century ago. One of them, a 1,200-square-foot image on the back of what is now a Verizon building, is dedicated to the Kumeyaay.

Toft grew up near Ramona and as an adult roamed the world as a professional wildlife photographer and tour guide. In 2005, at the age of forty-three, he returned home to stay. Toft moved into a new house surrounded by ten wild acres. "I loved it!" he recalls in the introduction to *Wild Ramona*, his impressive book of camera trap photos. "Starvation Mountain, with its house-sized boulders, was my view to the north. A strip of blue ocean was fifteen miles to the west."[87] On his land, Toft planted a garden of agaves, aloes, and palms, along with flowering plants to attract hummingbirds and butterflies.

He reveled in the nearby wildlife: hooded orioles, yellow-rumped warblers, roadrunners, red-tailed hawks, and lizards. At night, he listened to coyotes in the distance.

As a professional photographer, he had captured images of jaguars, caimans, cheetahs, and herds of elephants. But when he returned to California in 2005, he felt a special love for the animals on his land, though he seldom photographed them. Then on October 23, 2007, the Witch Fire and the fires in Bandy Canyon swept through, burned down his house, and wiped out much of the life on Starvation Mountain. Toft says the fire "wiped the slate clean." Soon, the land was impenetrable because of the overgrowth of laurel sumac, scrub oak, and ceanothus shrub, all of which tend to replace chaparral after a fire. Within a few years, however, some of the chaparral revived, and the valley gained a mile-long trail.

"COVID-19 hit just as my wife, Stella, and I were coming home from Botswana," Toft recalls. "No toilet paper, no contact with friends or family, no travel, and no work for me! All my photo expeditions and tours were canceled for the foreseeable future." By limiting Toft's horizon, the pandemic further opened his vision to the life immediately around his home. He set up camouflaged blinds in his backyard to photograph birds. And he invested in high-end remote-camera setups.

He describes his method:

Place a camera, generally with a wide-angle lens, low to the ground, and in a location where you hope to capture wildlife. I always try to find a pleasing background and consider how the sun throws light on the scene throughout the different times of the day. I carefully place three to five flash units around the location, set along the trail, to best light the subject and surroundings. Then I walk away and let the animals and birds trigger the cameras as they travel, forage, and drink or bathe.

He checks the cameras every two weeks, installing fresh batteries and memory cards. "Camera trapping is all about time in the field, understanding that 95 percent of images go directly into the trash can," he writes. "False triggers with nothing in-frame, animal butts, misfired flashes, or any number of mishaps can plague anyone doing remote-camera work. After weeks of nothing, I pull up a perfectly composed and well-lit bobcat looking back at the camera with the warm glow of sunset in the background."

Wild Ramona is a collection of three years of images from six cameras, representing 157,680 hours of "continuous camera operation." The photos are breathtaking. Unlike the typical trail-cam setup, which uses a single flash, Toft's cameras, surrounded by carefully placed nighttime flash devices, result in almost three-dimensional images of wildlife, including mountain lions.

For better shots, he rearranges boulders along a creek to create water holes, which serve as magnets for wildlife. With no human distraction, the photos reveal muscles that move like ocean waves, eyes filled with yellow consciousness, and even a hint of humor. Most affecting are the photos and videos of animals with their offspring, including cougar cubs nuzzling their mothers and suckling and purring.

Coyotes are the only animals consistently suspicious of the camera traps. They approach a camera, stare at it, and then dissolve into the bushes. Or they leave the trail to slip through the parallel brush. They go about their own lives, stay in their own worlds, and keep their own counsel.

THREE PHOTOGRAPHER FRIENDS occasionally drive to Julian. We head out along the roads and trails for what we like to think of as photo expeditions. I carry my iPhone. They think I should get what they call a real man's camera, maybe one like theirs, with a lens the length of a baseball bat. They do shoot stunning photos of wildlife — capturing bald eagles cavorting high over Lake Cuyamaca, dropping trout, then swooping down to grab them right before they hit the water. Either the eagles are conserving their energy, or they're playing with their food.

For now, I still prefer my iPhone. One reason is that it's always with me.

For a long time, after we moved to the Julian area, only broad landscapes and occasional animals (when I could get close to them) drew my lens, probably because of Ansel Adams.

I've installed several trail cams on our acre. As on Volcan, I periodically collect the memory chips from my cameras, insert them into my computer, and review the photos and videos. (Most trail cameras can shoot both.) In darkness or fog, I capture skunks, wild turkeys, owls, raptors, coyotes, bobcats, deer with wounds and rugged winter coats shedding in the spring, feral cats, a rare ringtail cat (a member of the raccoon family that looks like a weasel with a long banded tail), and an apparently pregnant opossum who seems to roll uphill. And occasionally, a mountain lion passes through, and I become fourteen again.

Recently, five foxes arrived, including three kits. They swoop in, tails like flags in a low wind. Small, quick, and eager, they glide over each other effortlessly. Their eyes are bright, full of extreme life. Dipping and turning, interweaving with the others, sometimes flipping in the air as they play, they exude fox joy. One stands up and rubs his whiskers and snout against the camera lens, then appears to do a backflip.

They drink meditatively from a bowl of water we refill every other day. Crows and ravens will bring a piece of bread to the bowl, drop it into the water, then pull it out and gulp it down. The foxes sip. Now and then, a fox will squat and pee beside the bowl or into it. The fox is saying, "This is mine." Occasionally, one of the foxes stops, looks off into the darkness, and freezes. It senses something in the dark. What?

The foxes live in a thicket next door. My trail camera recorded one of them running across a clearing carrying a squirrel. Suddenly, the fox stopped, dropped the dead squirrel on the ground, and then ran out of camera range. Seconds later, the fox returned, carrying another dead squirrel, dropped it next to the first squirrel, and looked around. And then the fox simultaneously picked up both squirrels in its jaws — one scoop. And trotted into the brush.

Three years ago, a mountain lion hid its kill in that same thicket. The carcass remains. Its rib cage looks like the framing of an old wooden boat, now gnawed on and stained by dried blood.

Just the other day, a bobcat popped up on the trail cam. It was a particularly startling night shot. What was it about this bobcat? Startled by the image on my computer screen, I stopped scrolling. Judging by the large

size of the animal, it was probably a male. This one was perhaps
pushing the thirty-pound limit
but somehow seemed larger
than his weight. Maybe it was
his shining eyes or his leaning-
into-the-night pose that gave him
such an aura of certitude and
strength.

What was he watching?
Maybe a covey of quail scuttling
up the dark hill.

Bobcats are usually too busy
dissolving into the high grass to give off this aura of authority. But other
people have seen the command of a bobcat at night. Here, the Viejas Band
of Kumeyaay Indians considers the bobcat "a cultural hero who first raised
them, and who first told the people the names of the months and divided
up the year," according to Ernie Cowan, a *San Diego Union-Tribune* columnist and nature photographer.[88]

Oregon's WildCat Ridge Sanctuary reports that some Indigenous people in the Northwest believed that "the bobcat and the coyote are linked
together in a theme of duality, with the bobcat representing fog and the
coyote representing wind."[89]

I'm skeptical. Coyotes are tricksters. Bobcats are cats; they would be
above all that. Hard to be mystical about a cat named Bob. But still…I'm
looking at that photo.

Every few months, my cameras will capture such an image. Sometimes
I understand why it moves me. Other times, I do not.

In one photo from our acre, a young doe stands at the forest edge,
watching the sun rise. Rays of light shoot through the branches and seem
to surround her. What is she feeling? What is she looking for? On another
early morning, a pair of turkeys and two young deer walk through a prism
of light. Two by two. Noah must be nearby. We want them on our ark.

Every wildlife photographer I know, whether amateur or professional, wants these animals to survive for a long time into the future. Each encounter allows us to see the world and our own lives through a different lens. First, the aperture must open.

In *Silent Spring*, Rachel Carson wrote, "One way to open your eyes is to ask yourself, 'What if I had never seen this before? What if I knew I would never see it again?'"

Or hear it.

CHAPTER 14

Songs of Nature

I N THE NATURAL world, we see with different eyes and hear with different ears.

I'm sitting at an outdoor café in Julian as a local man named Dave Johnston traverses the highway intersection at the end of town.

He leads with his white cane, past the Ramco Petroleum station. A truck stops to let him pass. People wave to him; one woman parks her car near the intersection and calls out, "Dave!" She gets out of her car and walks him toward his destination, the library. He waves his white stick at the cement. A few minutes later, he returns and walks past the coffee shop.

I step out onto the sidewalk and introduce myself. His smile is broad, his eyes opaque. We talk for a few moments. The man appears to be in his seventies. He steps a little unsteadily but does not falter. He's something of a legend in Julian. Unable to see in the conventional sense, he likes to hike up nearby Cuyamaca Peak, a nearby 6,515-foot mountain, sometimes at night. He usually hikes alone. He sticks to the road, reaching out with his cane, guided by the songs of nature.

"Today, all I hear are cars," he says. "Over at my trailer, it's getting louder and louder. Cars and trucks and people, I pretend it's all the sound of the ocean."

I ask him about his mountain hikes. Is he concerned about predators?

"A while back, a woman got killed by a mountain lion," he says, referring to Iris Kenna, a schoolteacher and dedicated bird-watcher. On December 10, 1994, Kenna was attacked as she hiked alone near Cuyamaca Peak. This was the second time a cougar had killed a human in California that year, after no killings during the previous eighty years. Investigators suggested that she had run from the lion, triggering its chase instinct, and that it had attacked her from behind.

That event, Johnston says, gave him pause. But it did not stop him.

"So, I got one of those sticks that electricity runs through," he says. "That got me through the fear period. But one day, the stick just fried out." Did he get another one? "No, no."

Johnston was up on that mountain just the other day. "Sometimes I went up there at night, but I won't do that much anymore. This one time, up there, a guy said to me that he'd seen a deer leg over there. Sure glad I wasn't there then. But I figure all the deer are fatter than me." He laughs. "I don't go on trails; I go on roads. And I don't camp overnight. That's not legal anyway, but I don't know who that would hurt."

Johnston is not unique or gifted with magical powers. Most of us see with our eyes but also with our bodies. Some people use a biological form of sonar called echolocation. Like bats, they bounce clicks and other sounds off surfaces to determine shapes and directions. This practice helps them navigate dark hallways or winding trails in mountain lion country. They also "see" by anticipating their next step and planning for the following millisecond. Will they fall? Will they encounter a bear? If their visual cortex can still process visual information, they unconsciously create a mental map of their location.

Nick Romero, author of *Sentience,* tells how, in the 1960s, Oxford University professor Lawrence Weiskrantz observed "a human patient, a partially blind man who was missing half his visual cortex, making consistently accurate guesses about the shape, position, and color of objects beyond his visual field. Weiskrantz named this ability 'blindsight.'"[90]

"What sounds do you listen for?" I ask Johnston.

He's short on time, he says. A pickup pulls up on the other side of the road; its front door opens, and a friend leans out, calls him, and says he'll give him a ride.

"Well, you've got to listen to the birds," Johnston tells me. "You can tell if it's day or night by the birds, partly. And they tell you if there's something wrong, to be worried about. In the morning, you hear those finches." He listens to the weather. If the air is warm on his skin, he knows it's daytime. Talking about this, his face fills with pleasure.

But what he really wants to talk about is the rush of people moving to the mountains, how it's making it hard for people to find places to rent. This is true in part because many of the houses have been converted into vacation rentals.

"The people moving here all say they love it up here because nature makes them feel insignificant! Well, I don't know about that," he says.

His friend calls across the road again, and he needs to go.

CURIOUS ABOUT HOW others experience nature's sounds, sights, and smells, I posted a question on Facebook. I invited people to share descriptions of the nature sounds they hear when outdoors or from inside the house, including animal sounds, plant sounds, weather sounds, people sounds, night sounds, and day sounds. These could be familiar sounds or strange, inexplicable, or disturbing. I was especially interested in sounds that hold personal meaning, offer comfort, or evoke a sense of mystery. Or music.

By the next morning, over 280 people had answered. Many comments were eloquent; collectively, they formed a kind of prose poem or aural image. As I went through the messages that day, my first impression was that these people were surprised — and excited — to recall or report so many distinct sensory experiences. Some of the comments came from our Julian neighbors. Others reached out from across the country, some from other continents.

Often, sounds were described in a musical context. Diana Umpierre, from Julian, wrote, "I hear the seagrape trees singing!" Umpierre is the former president of DarkSky International, a nonprofit that signs up cities and towns (including Julian) to reduce light pollution. "I knew trees could 'talk,' but I didn't know they could sing," she wrote. "The sound you hear is not birds (as I first thought). Pay attention: you'll hear the musical notes played by the branches and trunks."

The sound has a name. From the United Kingdom, Jake Ward wrote, "Psithurism is the posh name for the whisper of a breeze through a tree's leaves. At times, we can't hear the trees for the birds. Other times, the music of humans and the rest of nature is indistinguishable, at least at first."

Not surprisingly, birds were the most frequently mentioned sound source.

Danielle Bain, in Australia, offered this list of favorite bird sounds: "Gang-gangs croaking, crimson rosellas whistling, fairy wrens singing." Imagine Bob Dylan singing that. Gang-gangs, by the way, are cockatoos.

Jane Lynch, also writing from Australia, mentioned the laughing kookaburra (sometimes called the laughing jackass), the largest member of the kingfisher family. At home, kookaburras wake her before dawn, "just with their crazy laughs. They make me happy whenever I hear them." They remind her that "life is funny."

Growing up on her grandfather's Texas farm, Selena Beard Harris remembered hearing robins halt their "hellos to take shelter under the eaves of an outbuilding." Harris would watch from a barn loft as barn swallows "called me home," as she put it. She associated swallows with "the smell before a thunderstorm and the particular way the maple trees shifted and swayed; that feeling of being between two worlds."

Smell is the sense that most powerfully reminds us of earlier experiences. Sara Tsompanidi reports on a study by psychologist Silvia Álava, who "found that people remember 35 percent of what they smell but only 5 percent of what they see."[91] Sometimes we recall sound, scent, and sight as a multipart chorus.

Here's a suggested teaching opportunity from Amber Lilburn, who operates a forest school at her Florida farm. In her Facebook reply to my query, she described the sounds her students notice. "Lately, it has been a tie between the alligators and the bobwhite. An alligator sounds like an engine if you don't know what to listen for." Among the children, "that sound makes magical stories come alive."

While teaching preschoolers in Rock Creek Park in Washington, DC, Amy Garcia Bean noticed that her students "heard the mobbing calls of crows and were immediately alerted to the possibility of seeing a predator." Together, they followed the calls and "found the crows surrounding a great gray owl, a species not from the region, perched in front of a glass-covered building." They shared the sound with a local radio journalist who was hosting a bird expert at that time. On the air, the expert dismissed their claim; the great gray owl would have been "far beyond its territorial range." Her students sent photos of the owl to the expert.

He "had to 'eat crow,'" as Amy put it.

Some may see birding by ear as a new way to be a birder, but it's as old as the genetic structure that makes us hunters and gatherers.

"Just about every Indigenous culture practiced bird language where passerines (songbirds), were present," writes Lee Burton, founder of the Naturalist Training Program of the Catalina Island Conservancy off the California coast. Burton also teaches courses on animal behavior at Texas A&M University, the University of Florida, and the University of Idaho. On his excellent website, Burton further explains that "bird language is about both visual and auditory clues." And it's part of who we are: "As humans, [we] have a discrete bandwidth of super-sensitive hearing that's between 2.5 and 5 kilohertz, exactly matching the frequencies emitted by bird sounds (human vocalizations fall well below that range except for the 's' and high-pitched sounds). Many scientists now believe we evolved this capability, as it was essential for our own survival."[92]

Like our distant ancestors, people today are purpose-built to identify birds by sound. But the precise identification can be a challenging task. For example, placing a woodpecker is easy, but is it a downy woodpecker or its lookalike, the hairy woodpecker?

Your ears can help. Maybe. Woodpeckers don't peck so much as drum — the downy beats about seventeen beats per second; the hairy beats about twenty-six beats per second. According to the Cornell Lab of Ornithology, both birds beat so fast that you'll need spectrogram-recording software to tell the difference.[93] Over two hundred species of woodpeckers exist worldwide.[94] Each species drums at its unique speed and rhythm, which can change at any time to convey a message.

For multiple species living in the same forest, the din would seem to obstruct communication. But woodpeckers coordinate their drumming, stopping periodically so that every species has a chance to make a point.[95]

Imagine if your relatives did that during holiday dinners.

As the popularity of birding by ear increases, so do the American Birding Association's standards for identifying bird species. While the association makes no qualitative distinction between a bird heard and a bird seen live (not remotely), "a bird heard via a sound augmentation device may be counted only if the birder is present at the location of the device and hears the vocalization in real-time."[96] The National Audubon Society, tracking the rapid growth of birding, particularly after the pandemic, estimates that more than one in three US adults enjoys it, with the activity branching into new subsets. Some birders now practice *slow birding*, which encourages birders to take time to get to know the bird they hope to add to their life list of birds seen or heard.[97]

Why limit the tools of birding to birds? Why not create competitive events for other species? Froggers. Insecters. Skunkers!

Jeffrey Willius, author of *Under the Wild Ginger*, recalls a wilderness trip he took in Minnesota with five friends when he was fifteen. Sitting around their campfire, they "heard something that just didn't fit…a faint, rhythmic grinding, or munching sound" coming from near the edge of the woods. Willius followed the sound to a single pine tree. He shined his flashlight at a spot on the tree where the grinding and munching was happening.

"Aha! A BB-sized hole in the bark, and all around it a golden extrusion of hardening resin," he writes. "I put my ear as close to the rough bark as I could without touching the sap and listened for the mysterious source of the sound." When the trip was over, a camp naturalist informed him that the pine engraver beetle (*Ips pini*) "gnaws its way through the wood like a tiny beaver, just below the bark."

There, the naturalist said, "the adult male excavates — or engraves — a lattice of tunnels."

NATURE SOUNDS CAN HELP us heal and boost our empathy.

"The sound in nature that means the most to me is that of the morning dove," wrote LaVonne Misner, responding to my Facebook request. "While I was recuperating after nearly dying when I was only 24 years old, that's what I heard from my bedroom window. I'd received 14 pints of blood, and I could hear those doves doing their cooing, and all I could think was, 'I am alive.'"

Similarly, Carol Gulotta Sottili shared how, when she was six, she became seriously ill with scarlet fever: "I was bedridden for several weeks, sick, scared, and lonely. But a small, plain bird would sit in the tree by my window and sing its heart out for long periods each day. In my child's mind, I decided he was singing for me so that I would get well." Even now, at sixty-two, she always stops and gives thanks whenever a song sparrow sings.

When I shared some of these comments with journalist Dean Stahl, he said he realized just how much the songs of nature meant to him after losing his hearing. Antibiotics restored most of it, but he remained "profoundly deaf for several days." During that time, the sounds he heard mostly came from within his own body, which he discovered is a "noisy place," a one-person band of "whooshes, huffs, swishing of fluids, crackle of moving bone — that kind of thing." He found the sounds so disruptive that he had difficulty concentrating or relaxing. "I wondered at the time how it would be possible to live without hearing the external world. Wind on leaves means more to me now." Fortunately, he fully recovered. But the experience gave him a new appreciation for the auditory benefits of nature.

Listening deeply to nature also nurtures our ability to identify with other animals, to understand their joy and their pain. When one woman, from New Zealand, wants to skip her daily walk, she reminds herself that "this is their favorite time of their day too," and birds will have a shorter life than hers, so experiencing and appreciating their songs "is the least [she] can do for them."

Impressed and moved by many of the answers from the informal Facebook survey, I shared them with the ecopsychologist Patricia Hasbach. "I was struck by the intimacy shared in the posts," she said. "I wonder if sound is an often-overlooked part of intentional awareness. Too often, discussion of our senses is weighted toward vision."

She also pointed to a growing body of scientific evidence suggesting that soothing nature sounds may help us calm our sympathetic nervous system, which is usually activated when we're stressed or in danger. Nature sounds may also help reduce the body's negative response to stress, improve working memory and cognitive thinking, enhance our ability to focus, foster productive thinking, and help us avoid negative thoughts.

How you listen matters. Buddhist monk Thich Nhat Hanh as well as authors Jillian Pransky, Jessica Wolf, and Valarie Kaur, and others are proponents of what they call *deep listening* or *mindful listening*, sometimes using the terms interchangeably with *compassionate listening*.

In her book *See No Stranger*, civil rights activist and filmmaker Kaur describes deep listening to others as "an act of surrender...We risk being changed by what we hear."[98]

She provides a brief introduction to deep listening. Each of these elements can apply to experiencing nature.

"When I really want to hear another person's story," she writes, "I try to leave my preconceptions at the door and draw close to their telling." She quiets her thoughts and consciously listens with her senses. "Empathy is both cognitive and emotional." Understanding another person's view of the world requires experiencing the world with them. "The most critical part of listening is asking what is at stake for the other person...When the story is done, we must return to our skin, our own worldview, and notice how our visit has changed us. So I ask myself, what is this story demanding of me?"

* * *

NOW TO THE MUSIC of nature and how it shapes us.

Music was introduced to us by the natural world. Our earliest musical instruments included shells, parts of plants, and ram's horns. The ancient, tubular didgeridoo, a wind instrument linked to Aboriginal peoples of Northern Australia, is traditionally crafted from hardwoods hollowed out by termites. The termites are then collected for food.

The Dutch art professor Jan van Boeckel introduced me to joiking — or yoiking — a chanting that is similar to the music created by Siberian shamanistic cultures and that mimics sounds from the natural world. He mentions oral traditions suggesting that fairies and elves gave joiking to the ancient Sámi people, who lived in the coldest parts of Northern Europe. The Sámi would use their voices and musical instruments to essentially sing back to nature. Joiking is one of the oldest continuously performed musical forms in Europe.

The music of a yoik is not linear, with a clear beginning, middle, and end, but circular. Sometimes, it starts and ends suddenly and unpredictably. Nils-Aslak Valkeapää, a human rights advocate for his fellow Sámi, described the function of joiks as "not just music" but much broader: "They are paths to social contact. A way to calm reindeer. To scare wolves… The yoik was used to recall both friends and enemies. Land and areas. Animals. The yoik was also a means, a step. A step into another world. And this makes it religious… The yoik has no length; it never really starts nor ends."[99]

Van Boeckel says that Valkeapää compares the yoik to water flowing with the landscape or to wind brushing the mountain range. The art professor adds, "The yoik is like a ring circling in the air. Its rhythm is free, as are the rhythms of nature."

Whether ancient or modern, music can erase the boundaries between people and between humans and the rest of life on Earth. This is the soundscape of our shared existence on the planet. In his *Sand County Almanac,* Aldo Leopold, a key figure in environmental history, describes the call of sandhill cranes as "the trumpet in the orchestra of evolution."[100]

Composer Claude Debussy once said, "There is nothing more musical than a sunset." Bernie Krause, a music engineer and musician who introduced the Moog synthesizer to the Rolling Stones and the Doors, left his career in popular music to focus on the music of nature. Instead of working with rock stars, he records natural soundscapes on every continent, capturing them before they fade away. He is now a leader in the emerging field of soundscape ecology and the author of *The Great Animal Orchestra*.

"Some people say animals don't feel emotion," Krause once told writer Sarah Sweeney.[101] Then, she wrote, he played "a heartbreaking recording of a male beaver crying while he circled the river searching for his family, which had been blown up by dynamite earlier that day." Krause has observed the parallels and similarities between wild animals and other natural phenomena, including the electromagnetic frequencies of lightning storms that travel through our planet's crust. He has also noted the gradual disappearance of natural sounds. In 2023, he set up his equipment at a location where he had collected nature sounds for thirty years. As usual, he recorded for an hour. This time, he heard no sounds of nature.

Krause now fears that a truly silent spring may be approaching. The sounds and songs of nature are increasingly drowned out by leaf blowers, road graders, cars, and airplanes.

Garth Paine, an associate professor of digital sound and interactive media at Arizona State University, argues that climate change will decrease plant density, which in turn alters "the balance between absorptive surfaces, such as leaves, and reflective surfaces such as rocks and buildings. This will increase reverberation and make sound environments harsher" and add to already-disruptive sounds like truck engines. "In settings where sound reverberates for a long time, such as a cathedral, it can become tiring to carry on a conversation," Paine explains. "Increasing reverberation could have a similar effect in natural settings. Native species could struggle to hear mating calls. Predators could have difficulty detecting prey. Such impacts could cause populations to relocate, even if an area still offers plentiful food and shelter."[102]

The din of our culture is destructive to life, but we still have time to notice and be inspired by natural soundscapes.

"Nature is part of our humanity. It contains an imagination greater than ours," says celebrated cellist Yo-Yo Ma. In 2022, he participated in the Audubon Society's extraordinary Birdsong Project, which encouraged musicians to create hundreds of musical tracks to celebrate birdlife. Ma performed "In the Gale," written "for cello and birdsong" in a forest with a chorus of living birds.

In recent decades, scientists have paid more attention to the music of wildlife. Tokyo psychology professor Shigeru Watanabe discovered that Java sparrows are able to distinguish between harmonious music and discordant music, and they prefer the harmonious. Sanders advises us never to forget, "while oohing and aahing about the natural world, that none of it has evolved for us." And she warns, "If you're not watchful, the wonderful is made mundane." Sounds, like places, can become so familiar that we tune them out. "But on a good day, the mundane can be made miraculous."[103]

Of course, not all nature sounds are soothing or helpful.

In her Facebook response, Sarah Simmons described how, one hot night in East Texas, "the cicadas were buzzing so loud in the trees that I told my mom, 'It would be so creepy if they all stopped at once.' And then, they did." She added, "I felt like I was in a horror movie."

At our home, woodpeckers on the roof drop acorns in the ventilation pipes. The acorns clatter down the pipes and through the walls of the house. They sound like coins going down a drain. We need to look into this.

Still, the relative lack of *human* noise here has deepened my appreciation for sound. I've learned to recognize the birdlike warning calls of ground squirrels, the thumping of deer hooves, and the plaintive cry of a red-tailed hawk or the whistling of a bald eagle. Turkeys in the bushes drag their wings; the quick rush through high grass could be from a bird, a squirrel, or a snake.

One week, across the road from our house, workers from the phone company struggled to dig a hole for a new metal telephone pole. The workers hit rock. So they brought in a jackhammer. That morning, a helicopter flew over our house with the pole. The chopper hovered above the trees, making a noise like a Harley from Mars. The crew lowered the new metal pole, and then the chopper flew away, swinging the old wooden pole over our house and the treetops. As the loud beating of the air faded into the distance and finally stopped, I realized what I had been missing all week.

Not the quiet, but the music.

IV. BIOENCHANTED

en· chant· ment
1. *a feeling of great pleasure; delight. "the enchantment
of the mountains"*
2. *the state of being under a spell; magic. "a world of mystery
and enchantment"*

— *Oxford English Dictionary*

*This is what fascinates me most in existence: the peculiar necessity of
imagining what is, in fact, real.*

— PHILIP GOUREVITCH

CHAPTER 15

Otherworlds

I N THE NIGHT forest on Volcan Mountain, bats flittered across the sky.

In front of a screen outside the Volcan Mountain Foundation's education center, Ethan Staats, a marine biology PhD student at UC San Diego's Scripps Institution of Oceanography, where he is researching biofluorescence in marine organisms, conducting research in the affiliated Deheyn Lab. He was explaining the unusual ability of some plants, animals, and rocks to reflect light or produce their own.

Or at sea, to light up the waves and depths.

A few minutes later, a dozen of us picked up our packs and followed Staats up a treacherous trail, weaving through the forest.

We were wearing special glasses with yellow lenses. Periodically, Staats shone his black light into crevices, on the ground cover, and at the rocks of the forest. An astonishing otherworld burst forth, glowing ferns and rocks and mushrooms and even lit-up lizards.

He then pointed his black light at lichens growing on trees, the scaly leaves of incense cedar trees, and what might as well be fairy dust on the backs of scorpions.

Janice Bina-Smith, my neighbor and the nature therapist who introduced several of us to forest bathing, walked ahead of me.

"Listen, Janice," I said, "do you think these colors and lights could be what our ancestors interpreted as elves or fairy rings or spirits in the woods?"

"You're confusing bioluminescence and biofluorescence," she said.

She was right. The colorful lights we were seeing were the product of biofluorescence, a phenomenon in which an organism absorbs light at a specific wavelength and emits it at a different wavelength, which is then seen as a different colored, ultraviolet (UV) light.[104] Only animals with yellow lenses in their eyes can see UV light, like reindeer, butterflies, and sockeye salmon. However, because we lack such lenses, this kind of light is invisible to us unless we use black-light lamps or wear special yellow glasses. Our ancestors lacked such tools, making it unlikely that they could have interpreted biofluorescence as evidence for elves, spirits, or other mythological beings.

Still, some people whose eyes have been injured or have a genetic condition called tetrachromacy are capable of seeing lights and colors that others cannot see. In the past, people with one or the other of these conditions might have reasonably interpreted the colors of the world differently. And in the telling and retelling of these experiences, might their stories have become the imaginative stuff of fairy tales and other apparitions?

So, now let's turn to bio*lumin*escence, which is quite different from bio*fluor*escence.

Again, biofluorescence is the process of absorbing light at one wavelength and emitting it at a different wavelength. Bioluminescence, on the other hand, is more self-contained. A variety of organisms, including some deep-sea fish and terrestrial fungi, produce their own unique and visible light through biochemical reactions. That kind of light can be seen without special glasses or special flashlights.

As kids, many of us were enchanted by the blinking little bioluminescent insects (Lampyridae) that, in the Midwest, we called lightning bugs — also known as fireflies — hovering a few feet above the grass. (Back then, we didn't know that some female lightning bugs light their candles to lure males, party with them, and then devour them.) Our naked eyes can see this kind of light when ocean waves agitate bioluminescent plankton. We can also observe bioluminescence in particular species of squid beneath the waves, glowing roaches in the rainforests of South America, and bioluminescent glowworms that secrete glowing fluid from their anuses when disturbed. In addition, there is a shimmering, cyanide-laced species of millipede found only in the southern Sierra Nevada and a few other California mountain ranges.

But hold your millipedes, there's more.

Humans emit their own radiance, too. In 2009, Japanese scientists captured pictures of "glimmering" human bodies, the first images of human bioluminescence. The glow of the glitterati is usually too faint for the naked eye to see, but not always. The researchers reported that the light emitted by the human body can sometimes be visible to people and that its release follows a twenty-four-hour cycle, with the highest levels in late afternoon and the lowest late at night. They also found that human faces emit more light than their bodies, particularly around the mouth and cheeks.[105]

In addition to bioluminescence and biofluorescence, there's a third type of bodily light, called ultraweak photon emission (UPE), which is a natural by-product of metabolism. A new study has shown that most living beings, including plants, produce UPE throughout their lives. Daniel Oblak, a physicist at the University of Calgary and an author of the study, told *BBC Science Focus* that humans generate this UPE until the day they die.

Many of us who have taken a pet to the vet for the last time can recall that split second when the light in the pet's eyes goes out. Is that UPE?

I'm not suggesting that elves, ghosts, fairies, or other woodland spirits literally exist in the forest. But our senses do detect something in the trees, the dark corners of our backyards, or the city park—an animal, a plant, maybe even a human with that subtle glow.

Were Janice and I seeing forest spirits on Volcan Mountain on that luminescent night? Probably not. But we were enchanted. Or to be more specific, *bio*enchanted.

*   *   *

EVERY DAY, MORE COLOR is added to the canvas. A stream of new research continues to scramble our assumptions about communication, intelligence, and consciousness in the natural world. A Canadian research team reported examples of orca whales bringing gifts of food to people. Often, the whales wait until the humans react. Why? Possibly to study human reactions, or as a form of play, or because they believe we'll reciprocate?[106]

British researchers at Queen Mary University of London observed bumblebees that appeared to be playing with small wooden balls, pushing and spinning them. The researchers concluded that the behavior was not connected to mating or surviving. Bumblebees can also teach each other new tricks. They exhibited a "standing on the shoulders of giants' approach to learning," previously thought to be limited to humans, according to a 2024 *Nature* podcast report.[107]

Some bees also sing the body electric, harmonizing with flowers.

Until 2013, scientists were unsure whether bees could sense the electrical fields of flowering plants. Research now shows they do. As bees fly from flower to flower, they flap their wings to build up their own positive charge (up to two hundred volts). Bees then compare their electrical field with that of plants, helping them electrically identify their preferred plants. Pollen from the plant uses the charged field to jump to the bee, which is literally turned on. They use their electrical attraction for mutual benefit.

Electric pulses also help sea creatures hunt and avoid predators. The tiny elephantnose fish lives in freshwater lakes and rivers of Central Africa and West Africa. Researchers have revealed that it combines its electrical pulses with those of other elephantnose fish, essentially creating an electric warning system.[108]

To save lives and prevent infection, some ants will amputate their nest-mates' limbs (by biting them off).[109] Ants are also known to tend to the wounds of fellow ants by licking them clean.

But those are relatively minor skills compared with the talents of slime molds. *Harvard Magazine* asks, "Can slime molds think?"[110] Which makes you wonder what they think *about*. They can also, as reported by the University of Warwick, be kept as pets: "The slime mold *Physarum polycephalum* is very easy to keep, it's harmless and undemanding, it can live on a sheet of kitchen towel in an old margarine tub and needs just oats for food. Making maps and mazes for them to solve and explore is easy to do."[111]

Otherworlds even exist beneath our freeways and high-rises, deep within the earth.

Andy Murray, a photographer in New Zealand, wants to let the world know that an undiscovered otherworld exists in the soil beneath our feet, where, according to a report in the *Proceedings of the National Academy of Sciences,* over half of all life resides "including everything from microbes to mammals, making it the singular most biodiverse habitat on Earth."[112] The animals that Murray photographs range from cute to monstrous (in miniature). They're down there right now. Most of the time, we don't notice them unless we do what we did when we were kids: get on our hands and knees and inspect. Murray told CNN, "Every single person with a loupe can go into a garden or a park, and they can see the things that I see. It's all there — it's just another world."[113]

In November 2023, a *CNN Travel* report described a video documentary about underground life, a micro Jurassic Park: "An animal grazes, oblivious to another tracking it silently from behind. The predator waits for the perfect moment, then pounces. It's over in a flash." No dinosaurs required. The video reveals a battle between microscopic soil animals: a six-millimeter-long pseudoscorpion battling another ground-dwelling soil critter called a giant springtail, approximately seventeen millimeters long. The pseudoscorpion won.[114]

Consider, too, the phenomenon reported by Ferris Jabr in his book *Becoming Earth.*[115] He describes a portal in the middle of North America — in Lead, South Dakota — that reaches into "the deep recesses of Earth's rocky interior…through 370 miles of mining tunnels." There, Stanford University scientists have demonstrated that "contrary to long-held assumptions…Earth's interior is not barren," but populated with "intraterrestrial microbes…possibly living for millions of years. They often acquire energy in unusual ways, breathing rock instead of oxygen…Subsurface microbes carve vast caverns, concentrate minerals and precious metals, and regulate the global cycling of carbon and nutrients." Jabr adds, "Contrary to long-standing maxims, life has been a formidable geological force throughout Earth's history, often matching or surpassing the power of glaciers, earthquakes, and volcanoes... The history of life on Earth is the history of life's remaking Earth."[116]

Our ability to sense hidden nature is analogous to peripheral vision. As we approach an object, our eyes pick up more detail; each step forward reveals additional clues and opens up more possibilities. One way to understand this knowledge is through the analogy of peripheral vision, which shows a world that is usually half seen.

In his seminal book, *The Principles of Psychology*, William James writes, "Every one knows what attention is. It is the taking possession by the mind... of one out of what seem several simultaneously possible objects of trains of thought... It implies withdrawal from some things to deal effectively with others."[117]

But the French philosopher and political activist Simone Weil disagreed with James's insistence that attention requires the mind's concentration on the specific, while ignoring the rest. For Weil, tight focus was the opposite of true attention.

According to Weil's biographer Robert Zaretsky, she disapproved of such classroom mandates as memorization because they assumed there was only one right answer. "Reflecting upon a problem, rather than resolving it, was Weil's goal," he explains.[118] Her approach was akin to mindfulness, a way to receive the world while resisting judgment and avoiding overfocus. In other words, if you focus only on one tree, you'll miss the forest. And the otherworlds.

Bioenchantment and deep noticing work in tandem. Building on our imagination, noticing reveals the enchantment of the world. The more we notice, the more we notice. Bioenchantment illuminates the known and the unknown; it challenges us to discover the world as it truly is, not just a world reduced to numbers and measurements. Bioenchantment and noticing stimulate creativity and encourage and enrich our practices in both science and the arts by enabling us to ask better questions — to search for colors not yet known. To go deeper.

Today, I do more than walk. In the woods, I lift heavy slabs of black oak bark and inspect the patterns: overpasses and underpasses, freeways and canals engraved by industrious beetles. And I look out and up at dusk to watch clouds of bats and the insects they pursue.

At night, without the yellow goggles, Kathy and I see the others come forth. We wonder at waddling opossums and sashaying skunks waving their tails with no apparent fear; investigative coyotes who, like good journalists, wouldn't take their mother's word for it; and all the rest. Meanwhile, under the fallen leaves and the branches of ceanothus and manzanita, spiders march in a column toward a parade of black beetles. Add to these the pharaoh ants, thief ants, southern fire ants, and the industrious leafcutter ants who carry pieces of multicolored leaves like flags to their underground ranches, where they feed the leaves to their fungus herd. And all around, the mycorrhizal internet pulses. And the plants, all the whispering plants.

NADINA GALLE GREW UP in a suburb of Toronto, studied ecology and evolutionary biology, and later pursued a degree in earth sciences at the University of Amsterdam. She is now an ecological engineer whose work has been featured on the BBC. She was also named a National Geographic Explorer.

When Galle visited me in Julian, we walked up the same high trail I usually take to check the trail cameras on Volcan Mountain. As we moved through the forest, she said she believes we can utilize technology — what she calls "the internet of nature" — to help people in her profession plan urban forests and care for trees more effectively. High-resolution satellite imagery, for example, helps people gather environmental data, and AI software can analyze it.

We turned onto a side path through high grass to stand at an overlook. Looking out at the forest below, she talked about "the loneliness of trees."

While conducting research in Boston, she gathered nearly three hundred soil samples from beneath trees. "No tree had adequate fungi around its roots." They had been planted outside any natural system. These were trees that normally would be connected, exchanging information through fungi, possibly protecting each other. "But in Boston as well as Amsterdam, they were solitary, lonely creatures."

Was she anthropomorphizing trees? Probably. But she was using human language — words of empathy and compassion — to describe the lives of trees. She hopes to develop technological systems that can assist in the planning of urban forests and then monitor, sometimes from space, the fungal connections, water availability, and overall health of these forests — not just for the sake of the trees but also for our own benefit.

In this regard, seeing ourselves in a tree or a raccoon is a form of empathy, a way to witness another being, animal, or plant as part of us.

Not long after we hiked together on Volcan Mountain, Galle's book, *The Nature of Our Cities,* was published and will likely become a classic.

THE FORWARD EDGE of the life sciences keeps inching ahead of us, like agitated mycorrhizae.

Among the most eloquent modern chroniclers of the shifting scientific view is the previously mentioned Peter Wohlleben, author of *The Hidden Life of Trees.*[119] When I spoke with him in 2021, he described the plant world as far more complex than most of us consider.

In his classroom, he recommends that students look at a tree and imagine it standing on its head. Why? Because the perceptual life of the tree is in the base and roots.

He believes that the ends of these roots, which in some species are equipped with photoreceptor cells, may possess properties akin to a form of vision. This "vision" can help roots avoid rocks, for example, as they spread through the soil. He cites a study of *Boquila trifoliolata,* a climbing vine apparently able to change the shape of its leaves to match those of the tree it is climbing, presumably to camouflage itself.[120] We would suspect that the way the vine can detect the shape of the tree's leaf to mimic it would be chemical. But in this study, researchers planted a vine next to a plastic tree. As the vine climbed the fake tree, its leaves took on the shape of the plastic leaves.

Something other than chemical transference was going on here. Was this another form of seeing or feeling? Of sentience? Some critics suggested that the research methods were lacking.

Meanwhile, according to the BBC, researchers at Tel Aviv University announced that animals may react to the secret sounds made by plants, "opening up the possibility that an invisible ecosystem might exist between them."[121]

The researchers found that female moths avoid laying their eggs on tomato plants if the plants make a "distress" noise (that humans cannot hear). The researchers surmised that the sound may indicate that the plant is unhealthy.

"This is the first demonstration ever of an animal responding to sounds produced by a plant," Professor Yossi Yovel of Tel Aviv University told the BBC. "This is speculation at this stage, but it could be that all sorts of animals will make decisions based on the sounds they hear from plants, such as whether to pollinate or hide inside them or eat the plant."

This is the same team that, two years earlier, published research findings in the journal *Cell,* asserting that stressed plants emit airborne popping and clicking sounds, presumably as a sign of distress. Again, humans can't hear them. The team recorded ultrasonic sounds emitted by tobacco plants inside an acoustic chamber in a greenhouse. The sounds appeared to indicate whether the plants were dehydrated or physically damaged. The report suggested that the findings "may have a significant impact on agriculture."[122]

The possibility of this kind of plant communication raises questions. For example, if plants can send signals to moths, could they do the same with humans? We might imagine new technology that gauges the level of plant discomfort in a field of cucumbers and suggests whether more or less water or fertilizer is needed. The researchers are not claiming that plants are sentient, conscious, or aware. But, as Professor Lilach Hadany, also of Tel Aviv University, states, "This is a vast, unexplored field — an entire world waiting to be discovered."

SCIENCE MAY BE POISED at the beginning of a new research era that will confound what humans (or at least modern humans) have assumed for generations about cognition in the green world. Paco Calvo is one of the prominent biologists and philosophers now arguing that plants possess a form of intelligence and cognitive capabilities.

The director of the Minimal Intelligence Lab at the University of Murcia in Spain, Calvo is the author of the book *Planta Sapiens.* In an interview with science writer Amanda Gefter, Calvo says, "If the stuff that plants do deserves the label 'cognitive,' then so be it. Let's rethink our whole theoretical framework."[123] Calvo suggests that what he calls the *phytoneural system*—a plant's vascular network—functions similarly to our nervous system, conducting electrical impulses that may enable the plant to make decisions.

Plants may or may not know where they're going (though many plants move around), but they can sense where they are. They can, according to Calvo, discriminate between self and nonself, as well as between strangers and their relatives. They can anticipate, track the sun in the dark, exhibit entry-level sentience, and form symbiotic relationships with microorganisms in the soil. Twenty years ago, the case was closed: Plants cannot hear. That's probably not true, as it turns out. New science suggests that they can "hear" buzzing pollinators and the vibration of the wind.

"Plants reverberate both literally and figuratively," writes Caspar Henderson, in *Nautilus*. "Literally, there is a vine in the Cuban rainforest that has evolved bowl-shaped leaves that act as sound reflectors. The leaves help echolocating bats home in on the vine's flowers twice as fast as they do those of other plants, and in return for a drink of nectar, the bats pollinate the vine. Figuratively, the Japanese poet Matsuo Bashō imagines that he hears the sound of temple bells continuing in the flowers after the bells themselves have stopped."[124]

Multiple scholars, including professors of neurobiology and neurophysiology, have pushed back. In a paper titled "Debunking a Myth: Plant Consciousness" published by the National Library of Medicine, the authors conclude that, on several fronts, "claims for plant consciousness are highly speculative and lack sound scientific support." If plants are so smart, they argue, why don't they have concentrated clusters of neurons?[125]

Calvo's answer? Maybe intelligence does not require neurons or brains.

In the past, as Calvo puts it, "the mistake was to think that cognition was in the head. It belongs to the relationship between the organism and its environment."

"Plants are self-organizing, self-maintaining, self-regulating, highly adaptive," argues Evan Thompson, a philosopher at the University of British Columbia. "They engage in complex signaling among each other, within species and across species, and they do that within a framework of multicellularity that's different from animal life but exhibits all the same things: autonomy, intelligence, adaptivity, sense-making."[126]

The suggestion that some plants can communicate with and warn other plants about dangers gained traction in January 2024, when a group of Japanese scientists filmed plants doing just that. Using fluorescent imaging, the team, led by molecular biologist Masatsugu Toyota from Japan's Saitama University, gathered visual evidence of healthy plants sending airborne chemical defense signals to nearby plants after sensing volatile organic compounds produced by other plants responding to insect attacks or other forms of damage. The video recorder captured bursts of fluorescence spreading in a mustard plant, as if it were flushing internally.

A year later, students at Cornell University rigged a microscope to observe *Arabidopsis thaliana*, described by the *Cornell Chronicle* as "a small flowering plant widely used as a model organism in plant biology." Students in plant biology, computer science, and engineering teamed up to build a "working prototype...that detects a natural gene-encoded stress response in a plant to trigger a text message."[127]

The programmed text for this message was, "Hi, I'm hurt! Please help!" The effort was based on new research that reveals how "changes in pressure inside a plant's vasculature help transmit chemical and mechanical signals in response to stress," reports the *Chronicle*. The moment was described by a postdoctoral adviser as "both whimsical and historic."

The practical goal of such experiments by the Center for Research on Programmable Plant Systems (CROPPS) is to eventually devise ways for farmers (or large agricultural companies) to communicate with their crops, not in a casual manner, but for more productive and profitable fields.[128] "We're not just solving scientific puzzles—we're reimagining what agriculture could be," said Abe Stroock, director of CROPPS. "And sometimes that starts with a plant sending a text that says, 'Help.'"

The border between the virtual and the real appears to be dissolving.

Plant communication research suggests a larger truth: Life is defined not by objects alone but by the relationships between subjects. Without those relationships, do any of us exist? If a tree falls in the forest and no one is around to hear it, does it exist? Of course it does. One way or another, the other trees know.

CHAPTER 16

Interbeing

W**HAT YOU SENSE** depends on where you stand.

While walking around the side of our hill, I often stop to admire a ranch-style house. Gold-painted lion heads sit high on brick posts that bracket a gate, with Tibetan prayer flags flapping near two statues, one of Buddha and one (as I learned later) of a beautiful and motherly woman, the bodhisattva Jetsün Dölma.

Right now, I'm admiring not the lion heads but a cluster of household goods and tools piled in front of the gate. One of the items is a rolling desk chair. This is what Kathy would call "Rich bait." If I had a bumper sticker on my car, it would likely say "I brake for desk chairs and old typewriters."

Pushing my fingers into the blue fabric cushion, I wonder if this particular chair would go with my desk, also bought at a yard sale. A man in flowing robes appears outside the house and walks down to the gate and greets me. His black hair is wrapped into something resembling a bun. He is the Venerable Lama Lhanang Rinpoche.

"You may have this chair," he tells me. "The chair is free."

I test the cushion again.

Later, I googled him. He was born in Tibet and is a teacher of Vajrayana Buddhism of the Nyingma lineage, having studied with the Dalai Lama.

He directs a center for Tibetan spirituality, down the hill in San Diego. He is the coauthor of *The Tibetan Book of the Dead for Beginners: A Guide to Living and Dying,* which advises its readers on such subjects as karma — how we live is how we die, what dies and what gets reborn, our consciousness after death, and how to influence our next rebirth. His center on the California coast, Ngakpa International, offers in-depth education in "Chod, the Six Yogas and Dzogchen, Buddhism's most treasured path to cultivating clarity, presence and intrinsic wisdom," as the center's website explains.

Chöd is based on sutras explaining the Buddhist concept of emptiness. This concept is the belief that all things lack inherent existence. Instead, they're in a state of interbeing.

But I know none of this yet, as I stand in his front yard. Talk of the chair turns to other things, and he is an easy man to talk with. He smiles readily, with humor in his eyes. I like him immediately.

For about half an hour, we talk about nature, mindfulness, and — of all things — politics. (More about that later.) But mainly, we talk about how we see what we see. After a while, we turn to look out at the ocher fields and the dark, chalky green of the live oaks and black oaks sheltering birds and lizards. We take that in for a while.

"If you want peace, then be peaceful," he finally says, with his slight accent, still looking across the field. "If you want love, be love. How you focus and how you see depend on how you use your mind. Because in our mind, it is the ground of happiness and sadness. In the ground of everything. How you see the world, the world is the reflection of your mind." Our minds change, so does nature. When nature changes, so do our minds.

He pauses, and his eyes seem to look beyond the field and oaks.

"The nature nowadays and in our childhood is different," he says. "Maybe this is useful. When I was a child, I grew up in the northern Himalayan mountains in Tibet. So, there's no running water. There are no showers, televisions, computers, or electricity. So, we are just starting the day at sunrise. The sun goes down, have dinner, and then go to sleep. Very simple. People are very healthy."

I notice that the light is fading slightly over the field and through the trees.

He continues, describing how, when he was a child, his family acquired a clock and spent a lot of time watching it to learn how to tell time. "When we are starting to watch the clock, we want to know only the time. My father said, 'Why do we need a watch?' My father looked at the stars. He said that when this star joins with that star, it means it will be windy here tomorrow. Or, these clouds mean that tonight there is going to be rain or snow. Or wind."

This man doesn't see nature the way I do, at least not most of the time. I try to be in the present, in the here and now, to be mindful. That's good. But I sense that Rinpoche's here and now is both here and in eternity. The blackbirds that swoop under the branches of the tree in the field have been here before.

Having a chance for a do-over, a rebecoming, is a good thing.

ON THE OPPOSITE COAST, in the Hudson Valley, a remarkable young writer, poet, and ecofeminist named Sophie Strand says, "Myths are like palimpsests," older manuscripts physically obscured by more recent writing. "You might be reading the most recent text, but you can see glimmers below of older stories that are getting reappropriated."[129]

In her poetry, essays, and fiction, she describes a world in which biological and human mythic systems are inseparable, where the human body, mind, and soul extend beyond our skin or our brains, or what we usually consider our specific selves. She describes herself as "a neo-troubadour animist" raised by "house cats, puff balls, possums, raccoons, and an opinionated, crippled goose." Wherever she has lived, she "has been known as 'the walker.'" How does she define this moniker? In Strand's world, "all thinking happens interstitially — between beings, ideas, differences, mythical gradients." She also likes to say that her heart isn't in her body.

To Strand and increasingly to a growing network of scientists, this radical view of reality makes sense. Consider her description of the mind of a spider, which she roots both in science and her poetic and mythic sensibility. She describes a spider "seeing" by interpreting the vibrations that travel to it through its web: "Sitting like the iris inside a lacy eye, the spider tugs and flexes and tightens its grip on different strings…Each tug is a question, each returning vibration a reply…Where is the spider's mind? Is it inside the spider's actual brain? Is it in its spinnerets or legs? Is it in the web itself?"[130]

IN 2017, HILTON JAPYASSÚ, professor of biology at the Federal University of Bahia, Salvador, Brazil, suggested that "the web threads and configurations are integral parts of the cognitive systems" of spiders.[131] In 2020, *New Scientist* reported "an alien intelligence living among us," observing that some spiders "possess an extraordinary kind of consciousness, including minds that extend beyond their bodies" through the silk of their webs and that "some arachnids possess hidden cognitive abilities rivalling those of mammals and birds, including foresight and planning, complex learning and even the capacity to be surprised."[132]

Two years later, in 2022, *Science Daily* reported on the work of Ron Miles, a professor of mechanical engineering at the Thomas J. Watson College of Engineering and Applied Science at Binghamton University, New York. (Miles, incidentally, has revealed a lifelong fear of spiders, though that impulse may have changed.) His 2022 study of orb-weaving spiders, the species featured in the classic children's book *Charlotte's Web*, revealed that "spiders are using their webs as extended auditory arrays to capture sounds." In Miles's words, the spiders are "outsourcing their hearing" to their webs.[133] This bio-inspiration may help developers of extremely sensitive hearing aids.

That same year, *Mind Matters,* a publication of the Walter Bradley Center for Natural & Artificial Intelligence, suggested that cognitive as opposed to automatic responses "could be fairly common among spiders, who also exhibit talents of deception." The report's provocative headline: "Spiders Are Smart: Be Glad They Are Small."[134]

Spiders aren't the only creatures that think or perceive with extended brains. Octopuses are brainy-plus, too. Two-thirds of their neurons are located in their arms, and to a degree, those arms are capable of thinking and acting independently of, or in cooperation with, the central brain.

Alexander von Humboldt, a German geographer and naturalist who was, in the early nineteenth century, the most famous naturalist in the world, wrote, "Nature everywhere speaks to man in a voice that is familiar to his soul." He described the natural world as a living whole, "a web of life…a global force…interwoven with a thousand threads."[135]

Evolutionary biologist Lynn Margulis was on the same page as von Humboldt. In the 1960s, she proposed that parts of living cells, such as mitochondria and chloroplasts, are descended from once-independent bacteria. In the 1970s, with British chemist James Lovelock, she codeveloped the hypothesis that life on Earth is a self-regulating system. Some adherents to this theory contend that the planet — which they call Gaia — is itself alive and that life on Earth, on the brink of extinction several times, came to its own rescue. It achieved this through the sudden, intense growth of mycorrhizal fungi. The fungi helped the plants of the world absorb water and nutrients from the soil — and saved the day.

Lest the reader interpret this as a comforting solution to the climate emergency, these earthly self-corrections took millions of years to complete.

SUCH THINKING PARALLELS both the Buddhist view that consciousness is independent of the physical world, and the relatively new scientific arguments that the human brain does not confine consciousness and that the line between humans and other animals or plants is permeable.

Do we see our permeable relationship with the rest of nature as a good thing? Or would we rather pass our perception of nature to the machines — and avoid the messiness?

Looking deeper into this circular line of thinking, I learned that, in the 1980s, Thich Nhat Hanh introduced the word *interbeing*: "Western philosophy is preoccupied with questions of being and nonbeing, but Buddhism goes beyond the dualistic notions of being and nonbeing. I often say, 'To be or not to be, that is *no longer* the question. The question is one of interbeing.'"[136]

To illustrate interbeing, he suggests looking at a piece of paper. "If you are a poet, you will see clearly that there is a cloud floating in this sheet of paper. Without a cloud, there can be no rain; without rain, the trees cannot grow; and without trees, we cannot make paper. The cloud is essential for the paper to exist. If the cloud is not here, the sheet of paper cannot be here either. Looking even more deeply, we can see we are also in the paper."

The paper becomes, as he put it, "the object of our perception."

He added, "To be is to inter-be. You cannot just be by yourself alone; you have to inter-be with every other thing. This sheet of paper is, because everything else is."

WHAT DOES ALL THIS MEAN for our daily experiences in nature? When we walk, hike, or kayak, we can shift our minds from the name of a bird to the living whole, the life beyond its name, and the system it inhabits. This approach does not suggest ignoring the individual animal or plant, but that we enter the broader world that the bird and we share, first through imagination. While animals communicate in ways specific to their own species or community, they also communicate within and across species in the oldest universal language, which includes body movements, recognizable alarm sounds, and other signals that many animals make in their own unique ways.[137]

The concept of interbeing goes even beyond that shared language. It challenges the assumption that our minds stop at the skull and skin. It suggests a communion of consciousness among species.

So, are we humans or are we spiders? We sit in our webs waiting for some action. As humans, we think *we're* the action.

Sometimes when I walk the hidden paths of Volcan Mountain or down in the Anza-Borrego Desert, where late-afternoon light seems to emerge from the sand and rock, a ground squirrel will freeze and stare straight through me, or a fat tarantula crossing a road will stop both of us in our tracks.

One evening, I got down on my hands and knees in the middle of the road and conversed with Tarantino — that's what I named a furry, excitable tarantula I met as he and I were walking toward each other on the road above our house. I looked him in his eight eyes and talked to him. He seemed to cock his head, or rather his whole body, to stare back. And I wondered, Is my own consciousness, right now, being shaped by this spider?

Perhaps we attribute human characteristics to other living things because we believe we can relate to them, yet we know we cannot fully understand them. Is this a form of empathy? I'd like to think so. Anyway, maybe Tarantino was *arachnopo*morphizing me.

If so, we'll call it even.

The past three years have brought me closer to a spiritual view of nature — not *my* spirit in isolation, but the spirit to which I belong. Reaching that point does not mean adopting a recognized religion or belief system — unless mysterianism counts. In 2000, the prestigious journal *Nature Neuroscience* offered this description of the philosophical view, which "holds that even though there is nothing supernatural about how consciousness arises from neural activity, the human brain is simply not equipped to understand it." We didn't evolve with the cognitive abilities to directly understand consciousness, "just as armadillos did not evolve the ability to understand arithmetic."[138]

More optimistically, the editorial says, "The explanatory gap may not be closed at a single stroke, but it is getting narrower by the day." Or, a hopeful cynic might say that we'll solve the problem of neural consciousness around the time that armadillos learn long division.

Writing on *Medium*, David A. Palmer, an anthropologist with the Hong Kong Institute for the Humanities and Social Sciences and the University of Hong Kong, offers a succinct summary: "Throughout the ages, from tribal traditions to the scriptures of the great world religions, the purpose of the world of nature and its relationship with man has been a recurring theme. Regardless of the time and origin of these cultures and traditions, to varying degrees, the world is enchanted — it always has a sacred or spiritual significance."[139]

By contrast, "a materialistic view of nature places instrumental human desires at the center of concern. In this view, the natural world has no intrinsic value of its own; it is only valuable when it is used as resources to satisfy human needs."

Ecopsychology professor Tina Fields describes how we can turn away from this materialistic view: "We need not look far for an alternative worldview. In the Native American Lakota language, for example, the word *oyate* means 'people' or 'nation,' but is not limited to only humans; 'people' includes beings in other-than-human bodies, such as the "bird people/nation," and even more specific, the 'eagle-bird-nation.' In the Lakota worldview, humans are not seen as primary, dominant, or more valued than other life. Rather, we are considered weaker because we are a young species. In order to thrive we must humbly learn from our elders — other animals, plants, stones, and even weather patterns."[140]

Yet, as Fields points out, drawing too heavily from Indigenous cultures in the Americas risks cultural appropriation. Other continents and cultures offer their own wisdom traditions. Among them, the Celts who originated in Central Europe and spread to Britain, Ireland, and Spain. Celtic animism of the ancient world, persistent today, venerated the spirits of mountains, forests, watery places, and sky. Trees and their spirits were especially revered, as were the spirits of animals, including bears, horses, and crows.

One reason Buddhism is often more closely associated with the human connection to nature than, say, Christianity is that Buddhism places a strong emphasis on paying attention, being present, and cultivating mindfulness. However, a reverence for the natural world does run through Hinduism, Islam, Daoism, and Christianity.

Palmer writes that from a Daoist perspective, "a common life force circulates among all beings, linking them together in a single flow. Nurturing one's own life is inseparable from nurturing other life forms."[141]

Harold Coward, a Canadian scholar of bioethics and religious studies, explains that for the Hindu, "the universe is God's body, of which we humans, along with everything else in nature, are but a part."

This explanation sounds like panpsychism and related belief systems. "The essence of earth, air, water, the tree, cow, you, and me," Coward writes, "is the same divine spirit manifesting in different forms."[142]

Aviaja Lyberth Hauptmann, an Inuit microbiologist at the University of Greenland, seeks to synthesize the scientific method with traditional Indigenous knowledge of her ancestors through a holistic perspective on nature. "The Inuk word *sila* is not easily translated into English—it can mean weather, climate, spirit, wind, consciousness. What that points to is that it's all connected," she says. "We're not separate from nature; when our minds are well, when the sila inside of us is well, we can treat our environment well."[143]

THE NEXT TIME I SPOKE with Lama Lhanang Rinpoche, he was sitting on my deck. We were drinking tea that Kathy had brewed and were looking out at Volcan Mountain on a cloudless, windless day. Now we took up where we had left off, with politics — a topic that often ends conversations. Politics had started ours. He wanted to talk about how our radical and increasingly violent political polarization mirrors the human disconnection from the natural world. He began to tell me the story of how, during the Chinese invasion of Tibet, his parents were kidnapped by Chinese soldiers. Before their return, both were tortured.

He looked away. "This is why I am concerned now. People in this country think they are safe from such things. They are not."

Then he raised both of his arms and stretched his hands out to each side, like a tree or a cross. "Here is where we are now," he said. He wiggled the fingers on his right hand. "Over here, people talk and talk and argue among themselves." Then he wiggled his fingers on his left hand. "And over here, people talk and argue and point fingers at each other." Arms outstretched, he said, "What both hands have forgotten is they are attached to a body."

We finished our tea and went for a walk in our neighborhood. A deer, probably a doe, watched us from a dark grove of half-alive oaks and winding manzanita.

Rinpoche talked about birth.

"A lot of people say to my brother and me, 'Where are you guys born? Which hospital?' And I don't know. We have no birth certificates. My father always prepared the dirt for our birth. He would take out all the stones. When we were born, we didn't drop into cloth or fabric. We dropped and were born into that dirt. My father said that's very healthy, you are strong, you touch the earth."

I mentioned the scientific evidence suggesting that children who play in the dirt may develop stronger immune systems.

"So that's why I say we need the nature," he said. "Because we think, 'Oh, we are humans, they are animals.' But we are animals, too. We are one kind of monkey group, and thousands of years ago we lived like monkeys, and then we are getting technology and smart. And then we fight."

"Monkeys fight, too," I said. "Chimpanzees conduct their version of war."

But Rinpoche continued. "Then we invent swords, knives, bows, arrows, spears, try to kill each other. Then we think it's not enough. Okay, handguns, those guns, rifles. Then we think that is not enough, and we invent the machine guns. Then that's not enough; okay, we need a bigger bomb. And today, all the powerful countries have nuclear missiles. Who has the most? Who has the fastest missiles? Who has more intelligence? For what? All these we destroy ourselves. We're humans. We think we're smart, but we destroy our country's nature." We forget our interbeing.

A raven hopped and flew across the road. Rinpoche likes crows and other birds, especially ravens. They bring him news. Sometimes good news.

CHAPTER 17

When a Mountain Lives

ONE AFTERNOON, I drove a few miles west to the crossroads village of Santa Ysabel, California, just west of Volcan Mountain. There I met with Gabe Kitchen, a member of the tribal council of the Mesa Grande Band Diegueño of Mission Indians. We sat at a table near a window in a café. Two smartphones were on the table in front of him.

At forty-one, Gabe is the father of four children. Like many parents, he worries about how digital culture affects his kids and all the other forces pulling people away from the natural world. Concern about these issues drives much of his work to preserve as much open land as possible, including land that has been used for centuries for ceremonial gatherings.

"You know the same pressures are here," he said. "Pressure to build houses — and we need houses, or maybe a casino, though a casino is probably not going to happen."

Understandably, some of the pressure comes from within the tribe. "Someone sees one of these hillsides, and it's tempting to imagine some development going in," he said. "We do need housing. If we don't provide housing for a larger community here, then we're just continuing the cycle of disconnecting our people from our community and our culture."

Kitchen emphasized that he was speaking only for himself, not for the tribe or its members. But in a sense, he was speaking for all of us concerned about the integrity of land, place, and time.

He wonders whether there could be laws — in this region or beyond — that would encourage gentler use of the land. He mentioned smaller houses, for example. At least smaller than the mini-mansions that have carpeted Southern California's mesas over the past few decades. "Unfortunately, we've got to find someplace to destroy, to ruin, you know?"

Such problems are challenging to solve, no matter where we live. "There's no way we can ever go back to the way it was," Kitchen said. "With all the technology, it's just hard to find that balance. People want new phones. I've got two phones right here." He smiled and put the smartphones back in his pockets. "But we can make it better."

NEARBY, IN THE SANTA YSABEL region, the 1852 treaty between the United States and the Diegueño people was signed. (Diegueño is the name that the Spanish colonial powers assigned to the Kumeyaay in the eighteenth century. They are also known as the Ipai-Tipai and, in northern Mexico, as the Kumiai.)

The history of these people, spanning waves of colonization, has been turbulent. They were forcibly confined to ever smaller areas, and their freedom of movement was severely restricted. Traditional ceremonies and gatherings continued but were discouraged by the Spanish. Disease brought by colonists ravaged the Indigenous population, estimated to be as many as 20,000 in this region before contact with the Spanish, and reduced it to just 756 by 1910. Overgrazing, first by Spanish livestock and then by American, disrupted their ancient and ecologically sound farming methods. All of this and more has left a legacy of distrust and isolation.

Many San Diegans are either unaware that Indigenous people live in the county or unappreciative of the richness of their connection to land, water, and sky.

The Kumeyaay people were experts in resource management and environmental stewardship. To counter wildfires, they practiced controlled burning. They often lived in conical houses made of oak, willow, or sycamore branches and reeds. They migrated seasonally from mountains to the sea, where they gathered seafood in that radically different environment. And the Kumeyaay, then and now stargazers, assembled observatory posts in the mountains to watch and record astronomical events and occurrences in the night sky.

Angelina Lutz, an education specialist at the Imperial Valley Desert Museum (in Imperial County, adjacent to San Diego County), writes, "The Kumeyaay created star charts to trace celestial movements, mark solstices, and equinoxes. Kumeyaay observatories are large stone circles with spokes that indicate the sun's position at the solstices and equinoxes. They are usually built at high elevations."[144] Some of these circles are preexisting rock formations that still exist near the back trails of Southern California and Baja. On nearby Cowles Mountain, the Kumeyaay established a winter solstice observatory where they watched the rising sun split during the winter solstice, and they named the constellations: for example, Emuu (or Mountain Sheep), which corresponds to Orion's Belt. They called their collected sky knowledge *my uuyow*.

Today, the Mesa Grande Reservation is home to about 130 of the tribe's over 800 enrolled citizens, living in isolated homes, ranches, and small settlements. They farm, raise livestock, or commute to jobs in nearby cities and towns outside the 1,820-acre reservation.

FOR MUCH OF HIS EARLY CHILDHOOD, Gabe Kitchen lived on the reservation. At the time, he took his freedom for granted, but not now. Looking back, he values the inquisitiveness that roaming the mountainsides and valleys gave him. As a boy, his sense of the world was all-encompassing. He witnessed the movements of life as one system, the dirt and sky and water and wildlife. As a result, Kitchen could see the trees for the forest. He knew not only their species but also what he felt as their essence. He knew the familiar trees as individuals, each valuable in its own right.

What if those older ways were taught by Indigenous elders or other teachers in public and private schools today?

As a young man, Kitchen entered college in San Diego and studied kinesiology and anthropology. As a student in the city, he felt restless and unsettled. He later lived in Northern California and traveled through Central America, and in 2004, he returned to these mountains. Today, he works with the tribal government to preserve the land and water of Mesa Grande.

He had always planned to come home. And he suggests that what I call deep noticing was a way of life among his ancestors, though it may be fading in younger generations. He is determined not only to preserve the connection but also to keep it alive and grow it.

More than a place for humans to harvest fulfillment, land is an entity. It lives. In this context, deep noticing can be an act of respect for the land and the life it holds, as well as for the people, cultures, and traditions that came before.

"You should come to Tekamuk," Kitchen said. He was referring to the Tekamuk Gathering of Mesa Grande Band of Diegueño Mission Indians that would be held soon. "It's on Saturday. You should come."

A FEW DAYS LATER, I parked on the grass outside the Tekamuk grounds. A bald eagle landed in a high tree on the far hillside. The eagle looked down at the grassy bowl and hillsides of oak.

The early arrivals, including Kitchen, were setting up pop-up tent shelters for the Strong Hearted Native Women's Coalition and Empowering Native Wellness, along with feather banners (sometimes called sail flags) in front of most tents. One row of these tents offered coffee, burgers, potato twisters, funnel cakes, fry bread, and Indian tacos. Commercial tables were also arranged in a circle around a central field. One booth displayed knit hats and T-shirts, while others showcased handmade jewelry and art and other items for sale or on display.

Behind one table, without any written explanation, someone had hung a wall-sized sheet of plastic or vinyl. It bore an image of a giant oak, with faces of men and women peering from the high branches. The faces were smiling.

Waiting to sign the entry log, an older man sporting a white beard, a black ball cap, and a veteran's T-shirt ("One Flag, One Heart, One Nation") moved forward with the others. They wore black Western hats with beaded bands, some with beards, one with a long gray ponytail. Some of the women dressed traditionally, in long skirts decorated with ribbons. Many of the attendees were tattooed with images of celestial bodies, animal totems, and tribal symbols. Some of these images were traditional. A few men had facial tattoos of six- or eight-inch eagle feathers, one on each cheek.

I stopped to inspect a photo display of early Tekamuk gatherings, mainly from the 1930s. The names of the photographed older people, who wore traditional clothing covered with feathers, were memorable: Ramn LaChusa, Owl Clan; U'u (Horned Owl); Shechaak (Screech Owl), and Cinon Mataweer, who, according to the caption, "has become a symbol of solidarity and pride to the Mesa Grande Band. His life has taught us to value our history and respect our homelands. Although much was lost with his passing, his bloodline still exists and continues to reside in the traditional homeland."

At one tent, two men were teaching young boys the ancient art of flint knapping, chipping stone to make tools and arrowheads. The men and boys, serious and highly focused, used small antlers to chip the rocks. One of the teachers explained that this is the traditional method. "You can use copper, too." He pointed to several copper tools laid out on the ground. "But I consider that kind of cheating."

The day was young and people were still arriving, so the crowd was sparse, but that would change in the afternoon, especially when the games began. Among them were sack races and tug-of-war, with adults competing with adults, kids versus kids — twenty people on each end of a massive rope. My favorite was the truck tire roll. Starting at one end of the field, the contestants had to lift the tires into a vertical position (the tires were taller than some of the contestants) and then flip them end over end (not roll them) to the finish line.

Now, seven Bird Singers sat on folding chairs in front of a single microphone. One man, with tattooed cheeks and a graying goatee, was the lead singer. He held a rattle. All wore T-shirts; two wore ball caps, one a flat black fedora, and others were bareheaded. The ancient custom of Bird

Singing is a form of musical storytelling practiced mainly by the Kumeyaay and Cahuilla people in Southern California and parts of Arizona. A form of oral history, the songs and dances have been passed down for generations to tell the stories of the ancient islands off the coast, populated by the Kumeyaay before they moved inland. Using bird metaphors and allegories, the songs describe creation. One song, shared by the Cahuilla people in California, tells the story of a people who were lost until they watched birds migrating inland, flying as if one. The songs are instructive to children, illuminating the passage into adulthood, the need for children to fly from the nest, and how parents must let them go.

The Tekamuk singers began with a single shake of the rattle and then built from there. Kumeyaay singers do not use drums; instead, they communicate through the rasping rhythm of large rattles made from gourds or tortoise shells filled with seeds from native palm trees.

The men accompanied their performance with rhythmic chants and musical sounds unfamiliar to me, including transitions between songs marked by a short, low-decibel sound from deep within the singers.

For hundreds, perhaps thousands, of years, these songs have been performed at celebrations and ceremonies. Sometimes, during funerals, the Bird Singing continues all night and into the dawn. As the men sang, three women in traditional dresses moved forward. One of them was Fox Maxy, the video artist. They began to sway.

Later, I asked the lead singer why the three women stepped out like that.

"They were helping us," he said.

After the voices and rattles stopped, the lead singer stood to make an announcement: The ancestors had arrived. "They came to sit in the branches of the trees around us. They're listening now."

CHAPTER 18

Next

THE HUMAN RELATIONSHIP with the rest of nature is always in transition. For millennia, Indigenous people have created and experienced communion with the earth and lived with some semblance of harmony with the rest of life. Western civilization, not so much.

In Western cultures, the dominant attitude toward wilderness and natural landscapes has been ambivalence. For most of history, "wilderness was viewed as a place to fear and avoid," according to the Dutch researchers Agnes E. van den Berg and Cecil C. Konijnendijk. Nature was not only physically dangerous but also harmful to the mind and soul. "Even after the Middle Ages, Europeans abhorred the wilderness so much that travelers sometimes insisted on being blindfolded so they would not face the terror of untamed mountains and forests."[145]

From the late seventeenth to the end of the eighteenth century, during the European Enlightenment, scientific discoveries about natural phenomena led some Western intellectuals and wealthy city dwellers to see nature as the expression of an Intelligent Creator's will. During the Romantic period (roughly 1798–1837), nature "became the inspiration for the evolving concept of the sublime, i.e., a sense of awe and reverence, sometimes mixed with elements of fear," write Konijnendijk and van den

Berg. "The dominant poor rural population, however, still had to deal with the dangers of untamed wild lands."

In the mid-seventeenth century, French philosopher René Descartes introduced his theory of *bête machine* ("beast machine"). Contrary to viewing nonhuman animals as evidence of God's existence, he considered them to be mindless automatons lacking consciousness. The acceptance of animals as mere beasts suited the needs of the industrial age and helped justify all kinds of animal abuse carried out in the name of science and economic growth.

Today, one can see an echo or satire of Descartes in the kinetic sculptures created by Dutch artist Theo Jansen. His Strandbeests, as he calls them, look like giant Tinkertoy contraptions and seem to walk on a beach by themselves. Their movement is triggered when the wind shifts their balance. The Strandbeests look quite animated, but they're as alive as a sail on a ship.[146]

Now, the winds are shifting again, toward wide acceptance of the late biologist E. O. Wilson's biophilia hypothesis. Biophilia, derived from the Greek words *bios* ("life") and *philia* ("love"), describes the human tendency to seek connections with nature and other living beings. "To explore and affiliate with life is a deep and complicated process in mental development," Wilson wrote in his influential book *Biophilia*. "To an extent still undervalued in philosophy and religion, our existence depends on this propensity, our spirit is woven from it, hope rises on its currents."[147]

He also believed that science is converging with biophilia: "Modern biology has produced a genuinely new way of looking at the world that is incidentally congenial to the inner direction of biophilia. In other words, instinct is in this rare instance aligned with reason [and] to the degree that we come to understand other organisms, we will place a greater value on them, and on ourselves." Even so, some scientists resist the language of wonder. Their restraint is justified. They have their objective tasks to perform. However, University of California, Davis, historian Michael Saler writes, "Elites have enchanted themselves with the spell of disenchantment, but that spell appears to be breaking." Just when we think that enchanted banshees are banished for good, the return of the "magicians, mermaids, mesmerists, and a mélange of marvels once thought to have been exorcised by the rational and secular processes of modernity" occurs.[148]

As evidence of sentience in other-than-human animals and even plants continues to grow, a more nuanced and complex view of the natural world is spreading in Western society. In 2024, a group of prominent biologists, philosophers, and other scholars published and signed the New York Declaration on Animal Consciousness. The document describes strong scientific support for attributing conscious experience to other mammals and birds and "a realistic possibility" for aspects of consciousness in reptiles, insects, and mollusks.[149]

As of February 2005, more than five hundred scientists had signed the declaration.

In an article for *Slate* magazine, John Favini reports that scientists are "unearthing a densely collaborative world beneath our feet, radically shifting Western scientific thinking" away from the assumption that there are only two possible explanations for how life evolves: Darwin's survival of the fittest and creationism. Favini and others suggest a third way of understanding how life evolves: that collaboration and interdependence are more powerful forces in the shaping of life than previously thought.[150]

There is more to our lives than the reality we know. As a species, we've sensed this from the beginning. In fact, what tribal healers once taught is now revealed by science — not by proving that all myths of old are true but by uncovering a fuller reality.

When bioenchanted, we follow clues, half seen, heard, or felt, to a richer view of the natural world, even if we still have a few things to work out.

V. CAN WE MAKE PEACE WITH NATURE?

Once a young girl asked me, "Are the butterflies I feel in my belly part of biodiversity?"

— JOAQUÍN LEGUÍA

CHAPTER 19

Predators in Love

WITH HAIRPIN TURNS, Eagle Peak Road outside Julian winds along the edge of a deep canyon filled with craggy rocks and chaparral. Walking down this road one afternoon, I heard the squealing of tires behind me. A blue pickup truck driven by a young man with a ball cap and flying hair careened around a corner and shot toward me, hurtling down the hill in the middle of the road.

Before I stepped off the pavement, we made eye contact. Perhaps he could see my alarm. He hit the gas and shot past me. I could hear his tires squeal each time he came to one of the hard blind curves.

I had seen him before. He apparently had little regard for any life he might take, including his own. How could someone do that, repetitively, if they did not have a death wish? Taking out others—a deer, a fox, a child—would be of no consequence to him, in the spell of that wish. It's one thing to be careless about your own life but another to be thoughtless about others' lives. Is this behavior really a death wish or simply extreme egotism—lethal narcissism?

Wondering about this, I emailed Ron Swaisgood, the director of recovery ecology for the San Diego Zoo Wildlife Alliance. In this role, he oversees several species conservation programs in Southern California and around the globe. Ron, a good friend, was one of the people most responsible for pulling the giant panda back from extinction. I asked him if any animal other than humans will take unnecessary, foolish risks to themselves and others of their own species. Such hubris is distinctly human, right?

"Good question," said Ron. "I don't think there's an exact comparison, but there are some comparable behaviors. The main one is play. There's abundant evidence that play behavior can be risky — and on the surface unnecessary — but taking those risks helps the animal develop important skills to evade predators, compete with rivals."

Another example: animals that "show off" to gain social status or mates. This is known as the *handicap principle,* he said, such as the behavior of a bird called the babbler, which is known to approach and taunt snakes as a way of increasing its social position. But the one that seems most like your wild teenager is the young elephant who goes rogue, harasses other wildlife, and even kills rhinos. When this happens, wildlife managers bring in big territorial bull elephants, who put these teens back in their place.

So, when it comes to being jerks, we're not so exceptional after all.

A neighbor recently told me about how he had seen a careening truck scatter a gang of turkeys, then screech to a halt, turn around, roar back, and plow through the surviving turkeys, leaving more carcasses on the roadkill express. Most of us have witnessed such human cruelty.

Homo sapiens are the real king of the beasts. In 2023, South African researchers were surprised to find that elephants, giraffes, and rhinos are more afraid of people than any other apex predators; they run away from water holes 40 percent faster when they hear people there speaking calmly than they do when they hear a lion.[151]

It's unrealistic to expect us to change completely. But perhaps we can come to terms with our inner predator. That's what Kathy does when she carries ants out the door. As she passes, I suggest that the red and black ones in our bathtub may be fire ants. She arches an eyebrow and completes her task.

Back on Eagle Peak Road, I can hear coyotes. Two short yips down in the valley. A wild animal is mainly interested in efficiency, using only the amount of energy necessary to be fed and sheltered, to get the job done. Procreation, too. Yet coyotes, like our cats, play with their food. Two deer ahead of me leave the road and bound up the hill. Their fur is growing thicker and darker as the winter approaches. I don't know if hunting season has begun, but I've been told that when it does, the number of deer will increase on the hill. They hold close to the houses. Sometimes, men in camouflage move through the neighborhoods, where hunting is illegal, with rifles or bows. One woman reported on Facebook that she found arrows buried in the outer walls of her house.

Deer stop and watch me, seemingly without fear. The coyotes sound again below. I imagine their canine noses lifting, their ears back, their mouths opening, reaching for air and the others. During a recent visit to the nearby California Wolf Center, which works to preserve the Northern gray wolf and the Mexican gray wolf, I learned that both wolves and coyotes pull the same trick. They modulate their pitch to increase the illusion that their pack is larger than it is. Therefore, three coyotes (or wolves) can sound like a dozen.

Turning another corner, I watch wild turkeys waddle across the road. They stop to wait for a straggler. Usually, one turkey can't quite figure out how to get around or over the fence.

Then, I hear the swishing, thumping sound of wings, as the four turkey *vultures* fly over us in an arch and light onto the tops of telephone poles and into the high tree limbs. They make a thumping sound as they hit and hold the branches. They'll spend the dark hours up there, safe with their families and associates.

Ben Franklin is said to have wanted to make America's wild turkey the national bird. According to the Franklin Institute, that's a myth based on a letter that Franklin wrote to his daughter. There, he criticized "the original eagle design for the Great Seal." In the letter, Franklin wrote that the "Bald Eagle...is a Bird of bad moral Character. He does not get his Living honestly...[he] is too lazy to fish for himself." Franklin considered the wild turkey "a much more respectable Bird, and withal a true original

Native of America…He is besides, though a little vain & silly, a Bird of Courage."[152]

Were we to pick a new national bird today as a replacement for the bald eagle, a third symbolic option, neither predator nor prey, is available: the hardworking turkey vulture — the great recycler.

A week later, I was out with a group of friends walking in sweltering heat. We watched several turkey vultures circle above us, their heads tilted and looking down.

Wild turkeys and turkey vultures may look like cousins, but they're only distantly related. The impressiveness of the turkey vulture's flight is offset by its featherless red head and neck. The turkey vulture kills nothing (or almost nothing) and cleans up carcasses left by other animals, along with human garbage. Appropriately, the turkey vulture's scientific name is *Cathartes aura* ("purifying breeze"). A few vultures drop from the sky and join their compatriots on the telephone poles or high branches. (A group of vultures is called a *kettle* if they're in the air, riding the same thermal, a *wake* when the birds are feeding on a dead animal, and a *committee* — of course! — when resting in a tree or on the ground.)

As the group of friends stood looking at a gathering of turkey vultures on a house with a For Sale sign, my neighbor — an athletic woman who is a font of nature knowledge and who once biked across the Sahara — told us about a turkey vulture innovation. "When they're sitting up there and they get too hot, they cool down by defecating on their own legs," she said.

"I'll keep that in mind," I said.

Nature is beautiful but not always pretty.

IN OUR FOURTH YEAR HERE, Kathy and I drove down the western slope of the mountain to have dinner with Bill Toone and his wife, Sunni Black. As a young boy, Toone collected skunks. When skunks would visit his yard, he fed them. Carefully. After acclimating, a skunk would climb up and sit on his lap. Toone was never sprayed.

He has a way with animals, including humans.

While working at the San Diego Zoo's Wild Animal Park as a conservation biologist, he fell in love with a fellow employee. When he asked her out, she demurred. But Toone knew she loved butterflies. So, for two years, he worked to convince the zoo director to create a walk-in enclosure where visitors could pass through a cloud of butterflies. The Butterfly Jungle, as it was called, proved to be one of the zoo's most popular features. Sunni was entranced, too. Soon after the butterfly enclosure opened, they were married.

A persistent man, Toone also served on a team that helped save the state's largest vulture from extinction, the California condor. With fossils dating at least to the Pleistocene, the condor is the largest land bird in North America. This vulture, with a wingspan of nearly ten feet, can live for fifty years or more. In search of carrion, the bird can travel 160 miles a day. The California condor mates for life. It also blushes.

Until 1910, these condors were common in this county. They then disappeared, as they did throughout North America, because of habitat destruction, lead poisoning, DDT, and poaching, among other reasons. By 1987, the number of condors in the United States had plummeted to 27 birds. The survivors were captured and bred in the San Diego Wild Animal Park and the Los Angeles Zoo. One condor released in Mexico was seen on the California coast in 2007; it was the first condor seen here in more than a century. By 2022, worldwide population had grown to 561 birds.[153]

Toone was a main player in the condor comeback.[vi] For him, all of this began when he was a boy devoted to bug hunting.

I asked if he thought deep noticing could be taught. He said yes, then paused. When he was a grad student at UCSD, he said, he had been disappointed with a legendary biology professor. "He would only describe or see animals as objects, to be verbally dissected," he recalled. To be with animals fully, Toone said, requires temporarily unlearning some instilled habits of objectivity and distancing — not entirely, but balancing them with intuition and greater empathy.

In his book *On the Wings of the Condor*, Toone describes the excitement

---

vi   Toone, who lives in Escondido, California, is also a Fellow of the Royal Geographical Society and founder of ECOLIFE, a conservation and humanitarian nonprofit.

of catching his first turkey vulture for research. The bird "drifted down onto the ground near a carcass and trap." For nearly an hour, the vulture wandered around the dead animal body, inspecting it from a distance. Finally, the bird moved in for its first bite and then settled into its meal. Though the vulture was unaware of the nooses, "it was uncanny how successful he was at avoiding them."

Vultures gorge themselves. If frightened after a meal and too heavy to make a quick takeoff, they enact their backup plan: offloading a portion of roadkill flesh and offal.

"As soon as I saw a noose close around his leg, I rushed from my hiding spot to claim my prize," Toone recalled. The vulture covered him in putrid debris. "I nearly lightened my weight as well. But this was my first capture."

Smelly, damp, happy, and proud, Toone drove to his professor's home to share the good news. "Sadly, I have never felt so unwelcome in my life. The professor was a social animal, and he had guests that day for a late-afternoon get-together. I never made it past the front porch."

A few years later, Toone experienced another first encounter, this time with a giant condor. Hiking up a trail in known condor territory, he suddenly "felt a subtle shift in the air currents." The great bird, with a wingspan of ten or twelve feet, was gliding toward him. The feathers at the end of its wings tipped, and it came almost to a stop, turned its head, and looked at him. In that moment, Toone felt that he and the condor were connected…almost the same creature. As the condor rose on an air current, it turned its head again, still watching him.

Toone felt that he was seeing himself through the eyes of the condor.

EMPATHY IS THE ABILITY to see life from another's perspective. If we feel empathy for the caterpillar about to be eaten by a Swainson's hawk, can we also feel empathy for the hungry hawk or the plant that signals stress as we harvest it for our salad?

The more we observe and study any life-form, the more we understand it. And understanding fosters empathy. Or should. Yet, here we are, stuck in a war zone, the eternal dance between predator and prey. Intellectually, we know balance in nature depends on the dance. But cerebral understanding doesn't always make the dissonance go away.

John Irving, in his 1978 novel *The World According to Garp,* wrote about how Garp's young son mistook the word *undertow* for Under Toad, which grew in his mind into a massive amphibian that lived on the bottom of the ocean waiting to grab him by the legs. Thereafter, the Garp family used "Under Toad" to refer to the hidden disasters of life.

It's also a good term for the source of our ambivalence toward nature and its ways. During a recent walk, I stopped to watch four deer down a steep slope. They stood with their antennae ears up, watching me. I talked to them quietly, telling them that they were beautiful. They moved behind some trees and then came back out to stare at me again.

Now nervous, the deer trotted parallel to me, toward a tall wire fence. One after another, they glided, or levitated, over the fence. Up and down, vertical jumps, no sweat. Like young gymnasts. I heard the last one tap the top wire with its back hooves. They headed off across grass and through trees to another, higher fence next to a driveway. They levitated over that fence, too. All except the last deer, a small one, who faltered and missed the jump.

A dog, part pit bull, exploded from around the corner of the house, flailing spit and sound. We know him. He irritates us with his barking. The gentle woman who owns him is away most of the day. She is in her late seventies and her husband died years ago. The dog provides protection.

Deer can run forty miles per hour. Running away is their primary defense. The young deer spun and bounded across the property, back toward the first fence. The dog pursued it like a bullet, sinking his teeth into the deer's legs and flank, then locking into the side of the deer's neck. The deer ran, crying, dragging the dog. As they reached the corner of the fence, the dog pulled the deer to the ground. The dog was a ball of fury and glee, and the deer was wailing like a banshee, in heartrending squeals and cries. I ran down the hill along the fence, screaming at one point, "Barkley" — I'd forgotten the dog's real name — "Stop! Stop! Stop, goddamn it, stop!" The young deer struggled to its feet. Then the dog had it down on the ground again. The deer flailed, made that awful sound. The dog was on top of it, biting deeper into its neck, shaking the deer's head like a rag.

I slid and stumbled and leaped down to the bottom of the hill, finally within twenty feet of the dog and the deer. If I could have killed the dog with my voice, I would have.

Seeing or hearing me now, the dog twisted and ran. The deer, all spinning legs, flung itself at the fence. Once. Twice. It was still wailing. I looked. No spurting blood. Several red puncture wounds. A relief. But then I realized that the panicked deer felt trapped by me. So I ran back up the hill. I heard it hit the fence again. I turned to look. The deer was gone. I walked back down the hill slowly and could not see it. I heard no dog, no struggle. I presumed the deer had made it over the fence and into the woods.

The woman living on that property told us last year that a deer had gotten tangled in the fence while she was away, and it had hung in the wire for several days. She had to call a city department to remove the body. She said it was an expensive call.

If I had not interfered, I believe the dog would have killed the deer. But I had *already* interfered. I had rerouted the day. The four deer had jumped the fence to avoid me.

I finished the walk and went inside alone and opened a bottle of wine, something I seldom do.

EVERY BLESSING HAS A COST. The reintroduction of wolves in the West helps balance ecosystems. By thinning the sick and the slow from hooved herds, wolves protect the eroding banks of streams and even reshape rivers, sometimes returning them to their traditional routes. Mountain lions threaten remaining populations of bighorn sheep, even as the carcasses they stash provide food for countless mammals, birds, insects, and, eventually, plants.

The naturalist Joe Hutto demonstrates the paradox of attachment to other species. As part of his research, he raised a gang of wild turkeys from eggs. The chicks imprinted on him. To study wild turkey behavior, he followed them for months. He watched them eat and learned their travel patterns and habits, what they knew instinctively, and what they learned on their own. Over time, he felt he had almost become a turkey. Once, he found himself eating a caterpillar without thinking. All in the name of science.

Later, he bonded with mule deer in Montana and wrote another book about what he had learned as he traveled with them.

In 2017, *Sun* magazine published a long and fascinating interview with Hutto.[154] It focused on how he had become deeply attached to these animals as individuals. This type of research, he said, did not compromise his objectivity. "The more attachment you can develop with an animal, the more insight you get into its nature."

When his favorite turkeys or deer died or were killed, often at the hands of other human beings, he grieved. "I've been surprised again and again to see them survive some horrendous injuries."

Reading this, I thought of the number of times I have watched mule deer crossing our property, limping, with slashes on their flanks. I had assumed that those wounds were suffered during rutting season or from mountain lion or dog attacks. Barbed wire might also have been the reason for the wounds. Most deer, according to Hutto, "go no longer than a month without receiving some sort of injury from a barbed-wire fence."

In Montana, Hutto became especially fond of a buck he had known since its birth. The deer's name was Babe, and Hutto witnessed Babe's rise in the herd's ranks to dominance. "He liked me. We had good chemistry and enjoyed being together," he said. A poacher killed Babe. "I have difficulty talking about it to this day."

Since then, when he looks at a newborn fawn, he thinks, "Here's another tragedy waiting to happen. I still have dreams about mule deer every night." Over time, his feelings began to overwhelm him. When his wife died, he decided he had experienced enough loss of life.

Now came the *Sun* editor's hardest question: "Did your prior experience with the deaths of animals inform your experience with your wife and give you some understanding or acceptance?"

Hutto's answer: "No, just the opposite. I have not learned any extraordinary wisdom about the nature of life or come to see nature as kind. It's not. My wife contracted a terminal form of cancer. She had led a healthy life, never had a bad habit, and yet this awful disease took her…She was one more beautiful creature who died with her head in my lap."

He added, "I'm not accepting of the fact that we live in a profoundly brutal world. I don't fully approve of the way nature works. This lifetime of study has left me disappointed by the brutality of it all."

I was moved by the honesty and clarity of his answer. Sure, there's the circle of life and all that. We've been taught that objectively distancing ourselves is a virtue. Yes, we can see the fact of brutality as part of nature's way. But as Hutto suggests, we don't have to like it.

CHAPTER 20

Stories of the Dust

L AND SETTLES INTO US. If we stick around long enough.

In addition to being the most biologically diverse county in the lower forty-eight states, San Diego is home to more small farms and organic producers than any other county in the United States. The amount of agricultural land in use decreases each year, but since Spanish colonization, the backcountry has supported cattle ranching.

A few years ago, I visited one of the oldest ranches in the county, just to the west of Volcan Mountain. Our former San Diego landlord, a semi-retired professor from Loma Linda University, operated the ranch. After retirement, he and his wife moved to Mesa Grande, just south of Volcan Mountain. Cattle have roamed relatively free there for at least two hundred years. The ranch was anything but industrial. Lethal, yes, but a place where personal compassion for a farm animal and a connection to nature were still possible.

One night, I was standing with the professor's dog out under the river of stars, watching satellites and listening to the call of coyotes. They were moving closer.

When the professor said something quietly to his dog, it padded back up to the house. From a high plateau, we could see dark hollows lined with live oaks, the native trees that live so long and change so little that they don't really seem alive. We could see the black ridges of Mexico to the south, and only a handful of human lights in sight. "We thought we would nestle down into this," he said, as a satellite blinked above Mexico.

"The house on this place reminded us of our farmhouse back in the Midwest," he said quietly. "The kids are grown and we're here and Mesa Grande still doesn't seem quite real." He shrugged. "Maybe something happens when you get older. You don't think of anything being permanent anymore."

The next morning, he was up before dawn, frying eggs. The mountains were still black, but the sky was fired up. The professor walked across the gravel to his old Mustang, and in the quiet, his footsteps sounded like small explosions. He drove down a one-lane dirt road through the trees to the bunkhouse. As the sky lightened, he and the two ranch hands, Edmundo and Pablo, threw a bale of hay into the back of an old jeep and set out to gather the cattle.

Pablo, from the Mexican state of Jalisco, was twenty-one and wore his shirt open and a baseball cap on backward. Edmundo, forty-nine, had thick, graying hair and a gentle, almost delicate voice. Edmundo had come over the mountains from Tecate, Mexico, twenty years before, walking for nearly a week, night and day, and has worked on Mesa Grande ever since. He is as strong and dependable as ranch hands come, but sometimes his son from Tecate shows up and the two of them disappear for days of sad celebration. One time, the professor thought that Edmundo was dead or gone forever, but he had only been on a long-lost weekend. The following Monday, a rusted Cadillac filled with Edmundo's friends from Chula Vista had come roaring up to the mesa to deliver Edmundo and to celebrate his return.

ONE MORNING, I RODE in a jeep with Edmundo and the professor as they scouted for stray cattle. Edmundo stopped the jeep on a high ridge and squinted.

Dark shapes moved on a distant hillside. He cupped his hands around his mouth and made a high-pitched sound: "Woooooo!" And the professor called, "Sooo-kay, Rosie." It was a call that the professor's wife had used as a girl in the Blue Ridge Mountains. He called again, this time with one he had used as a Michigan farm boy: "Oh, oh, okay, Bucko." The dark shapes began to move down the hillside toward the jeep.

Edmundo and Pablo often see mountain lion tracks and feel as if they are being watched.

Milt Angel and Alphonso Soto were waiting for us at the branding corral. Angel is a big and wealthy man who seems younger than his sixty-odd years and poorer. He wore a soiled baseball cap with the CAT trucker insignia, droopy jeans, a denim shirt, and a layer of dust. He made his money as a builder in San Diego, then came home to Angel Mountain, where his family had settled over a century earlier. His great-grandfather had been among the first to cross Donner Pass after the tragedy of the earlier Donner Party.

Angel wiped wet dust from his forehead. "Can't remember a year this dry since I was a kid," he said. When he talked, he looked off across the clearing, but seldom directly at the listener. "My father had an old steel plow that year and I had to sit on the tongs to make them dig down into the earth. It's cyclical. We're due for a big snow. I remember standing in snow as high as this fence, and the roof creaking with the weight of it."

He paused and listened to a stirring in the oaks.

"You've never seen anything like this place, have you?" he said. "My father was born on Angel Mountain; I grew up in a house that stood right over there. People loved it so much they stayed a lifetime, not moving from place to place like people do today. Except for some of the second and third generation. They've started to sell." Clearly, this was Angel's mountain.

Up in the oaks, coyotes chattered suddenly, and the cattle began to come in from three directions, mewing and bawling and irritated. Pablo and the professor were out of the jeep and running behind a group of cattle, arms spread.

As the professor approached, Angel seemed agitated. "Those old cows need some supplement. Couple tons of molasses, in blocks."

The four men spread out behind the cattle and moved them into the branding corral. Angel and Soto, a Mission Indian he grew up with, went into the corral with the cattle and began to count them. Although these were not his cattle, Angel knew each by its markings. This one should go to the butcher today; that one, because it was carrying a calf, should not. In the heat and dust, Angel and Soto moved together, as they had since childhood, speaking half Spanish, half a code all their own.

"Damn," said Angel. "Good cow's gone, maybe lost."

The professor and Edmundo went in the jeep across the fields to distant corners of the ranch. Edmundo is a "good cow man," said the professor. "He can find them in the brush when no one else can."

Edmundo worried that the missing one had been bitten by a rattle-snake. "Three, maybe four days, muerta," he said, and folded his hands beneath his cheek as if going to sleep. The cow would eventually be butchered, but for now it was the object of compassion.

Edmundo left the jeep behind and set out on foot, watching the dust in front of him. He was searching for tracks. Cattle tracks. And, especially, mountain lion tracks. He spoke Spanish: "Here is a deer track. A big buck came by last night, maybe midnight. And here is the track of a fawn, six, seven, days old, here just this morning." He shook his head. "Ninguna vaca," he said. No cow. He walked out on a ridge above a long, deep valley that leads toward the sea. A fog bank lay in the valley, and a single mountain peak spiked through the fog. He stood quietly for a moment and looked at the valley, and then he sniffed the air.

"Muerta," he said again, and made his sleeping sign. But the wind shifted, and he was not sure. He stood there for a moment more and then turned and headed back toward the jeep and the professor.

NEARBY RANCHERS — or, more often, absentee landowners — now make halfhearted plans to cut the backcountry into subdivisions, flanks and loins of land to be consumed someday by RV parks and retirement condos. But for now, the land of Mesa Grande is alive.

Not far from here is a spring called Mem-sha-wee; drink from it, the Kumeyaay believed, and you will always want to return there.[155] The professor and his wife moved to Oregon a few years ago. They have not returned. Meanwhile, in most of the United States, small farms — except for organic enterprises — have all but disappeared, replaced by industrial agriculture.

In the Southwest, Edmundo and Pablo were among the remnants of an older culture. I wonder if they've been tracked down, rounded up, and deported. I wonder where they are now.

Below our house, just off the road, a small depression serves as an extra parking place. After a rain, a large puddle forms there, and as the water drains, the puddle turns into slimy mud. Sometimes, I'll walk down to look for tracks in the mud, and I usually find them. Some of the tracks are the size of my hand and show a three-lobed pad. Cat paws, including those of mountain lions, have two lobes on the front edge of the heel pad and three lobes on the back edge. Canine pads have one lobe in front and two lobes in the rear. While canine tracks usually show claw marks, feline tracks seldom do, because they have retractable claws.

Nearly a decade ago, Kathy and I drove to the mountains to take a tracking course focused on locating and reporting invasive feral hogs. Usually, these are domestic pigs that have gone wild, like werewolves with no full-moon requirement. Thick fur covers their bodies, and their eyeteeth grow into large, dangerous tusks. Feral hogs run in packs. Their mood is not the best. They displace and kill wildlife. This county's eradication program used tracking, trapping, hunting, and sometimes hunting with wild dogs. The program has been successful.

Kathy and I never saw a single feral hog track.

Still, I wanted to learn more about tracking as a way of noticing wildlife. My interest was further piqued by Jessica Carew Kraft, a proponent of rewilding ourselves as a way of life. In her book, *Why We Need to Be Wild,* Kraft writes that, "Tracking is akin to learning to read in a new language." It expands our perception and dismantles "our touristic mode of viewing the land around the trail as one big picture window for our aesthetic pleasure." Instead, tracking is a way to see "millions of stories written in everything." It prompts attention to the ground you walk on, the tracks that you make, the disturbance that you create in the environment.[156]

In his book, *Becoming Animal,* David Abram writes, "One of the countless signs that our thinking minds have grown estranged from the intelligence of our sensing bodies, is that today a great many people seem to believe that shadows are flat."[157]

Recently, my son Matthew and I went for a walk on an isolated segment of the Pacific Crest Trail. Tracks were initially difficult to find. But as the sun crept toward the horizon and shadows lengthened, the tracks stood out in relief. Now able to see the tracks clearly, we followed a highway of footprints, left by coyotes, bobcats, turkeys, raccoons, lions, and people. So many beings passing, pausing, moving on.

I wondered, What would it feel like to be tracked?

SOME PEOPLE READ the future in tea leaves. Ab Taylor found it in dirt.

I first met Taylor a decade ago, a few miles from where Kathy and I now live.

"Yessir, there's a story there, if you can learn to read it," he said one day in his soft West Texas accent. "Whether it's written by storm or animal or man." By studying dirt or leaves or mud, Taylor could tell where people have been and where they were going, and sometimes he could see how they had met their end.

He was a tall man with long, thick gray hair and a wide mustache curled slightly upward. In 1980, Charles Bronson portrayed him in a movie called *Borderline,* but in real life, Taylor didn't look like Bronson. He wore denim shorts and slip-on sneakers, a striped sport shirt with the tail out, and an alarmingly orange ball cap advertising "Ab Taylor's Step-By-Step Man Tracking."

In the 1960s, Taylor developed the US Border Patrol's first training methodology for trackers. One of his students, Joel Hardin, who became Taylor's counterpart in the Northwest, tracked down the Bulgarian Mountain Man, as the residents of Washington's Whitehorse Mountain came to call an escaped murderer who had found his way to the United States and then disappeared into the forest and evaded law enforcement for a dozen years. Hardin, reading trail marks as much as five years old, found the sixty-eight-year-old man in his final lair, a six-by-eight-foot hole covered by logs. The hole was so shallow that the wild man could not sit up straight.

During his career, Ab Taylor tracked countless migrants on their way north from the border. At the time I met him, I had recently written a long series for the *San Diego Union-Tribune* about the Mexican migration. The more I learned about its history and the hypocrisy and cynicism of both governments, as well as US border policies, the more I sympathized with the migrants.

Still, I liked Taylor, who believed in doing a job right, especially when he tracked families dumped off in the high desert by coyotes, as the often-predatory guides are called. The Mexican cartels control most of the coyotes now.

On the day I met him, Taylor was hiking to create tracks through the new Crestridge Ecological Reserve, a 2,600-acre mountainous parcel not far from Julian. Later in the day, he would send young men from the Urban Corps out to follow his tracks, to teach them his science. He walked steadily beneath the oaks, through the native mule fat bush (aka seep-willow) and sugar bush, carrying a spiral walking stick made of Tennessee maple. He explained that the stick was cut from a little tree that had been choked, stunted, and hardened by winding wild honeysuckle. "This little tree kept living, but it didn't prosper," he said.

He recalled tracking one group of Mexican Oaxacans for sixty miles, step by step, for days, from Tecate to Mount Palomar, west of Volcan, "slipping up and down the mountains — always, always at the same speed." He spoke with the respect that sometimes connects the searcher and the sought. "They had walked all their lives. Some of them had never ridden in a vehicle. Especially in the 1950s, when we transported them in the trucks, they would immediately become carsick."

By the 1990s, the Border Patrol placed increasing emphasis on sophisticated technology. Taylor had nothing against technology. Inexpensive digital cameras — and now cell phone cameras — were proving increasingly valuable, he pointed out. "You can take a photo of a track, enhance it on the spot to better detect the grid of a sole. You can print out copies for the searchers right there or transmit the image electronically to teams in the area, so that everyone is looking for the same footprint." Nonetheless, he believed that tracking ultimately relies more on human senses and deductive reasoning than on megahertz or fancy optics.

As the years went by, he switched from Border Patrol work to training search and rescue teams. And he volunteered for other causes, such as the futures of the dozen green-uniformed young men, aged eighteen to twenty-five, who had led difficult lives. They attended a special Urban Corps school to earn a high school diploma. With Harden, they were learning how to be trackers with search and rescue teams. The young men practiced two or three days a week, several hours a day, for six months.

They gathered around him now—one from Guatemala, others who had grown up in Mexico—poking and teasing each other; one or two stood still and quiet, their eyes distant. Later, Taylor speculated that he, or other men and women he had trained, might have tracked some of these young men's fathers or grandfathers. But in the moment, he smiled, eyes sharpening, and he began to tell them the stories of the dust.

Since then, how many of these young men have disappeared?

Today, if you drive into any of the busy little towns along the border, you'll see the long snake of old and new thirty-foot-tall steel bollard wall that runs from the Pacific Ocean, on and off, and along the Texas border with Mexico.

Parts of the wall were built higher during the first Trump administration, but at any height, the wall has been and remains a disaster for people and wildlife. The Center for Biological Diversity reports that ninety-six species are threatened by its construction. Meanwhile, not far from where we live, men and women still die in the desert, and their children are held in cages. Like wild animals.

In 2013, Taylor died at eighty-eight. In his last years, he taught adults how to search for lost children. And he taught children how to survive in the wilderness.

*　*　*

THE OTHER DAY, walking west on Eagle Peak Road, I passed an area where the hill had been cut away to make space for the road. In my peripheral vision, I saw the face of the rock, crushed granite, and light brown clay, with the ends of the rocks exposed to the air and showing new revelations after a recent rain. And I had the distinct feeling that this bank, this slice, one that looked like an open-faced beef sandwich, was in fact alive.

Surely my imagination was acting up. But there was such personality bursting from deep in the hillside, where it was cut clean to make way for the road. I could see the stories of dust storms and drought and massive floods and fire.

During moments like this, it's difficult to think of dirt and rock and dead branches as debris. Rather, they feel like evidence of breath.

How are we really attached to land? Especially this land of migrants pulled north and pushed south again with the political tides, now caught in the fifth great wave of deportations? The land stolen or adopted or borrowed by other immigrants? That land? Today, the land is moving again. As part of a rising Land Back movement in the United States, Canada, Mexico, and Australia, Kumeyaay tribal leaders are in the preliminary stage of taking back control of a large parcel of land on the north side of Volcan Mountain.

Land is not given back to but bought back by the tribes, alone or partnering with national or local conservation organizations, in this case possibly with the Volcan Mountain Foundation. The terminology is difficult, because how can land be sold back now if it was stolen in the first place and then passed down from generation to generation?

Beginning in the Paleolithic, more than nine thousand years before the Great Pyramid of Giza was built, some six hundred generations of Kumeyaay, Payómkawichum, Cupeño, and Cahuilla have lived here.[158] Then came the others. Spanish, German, Scottish, Russian, and Kansan. Their voices, too, can be heard in that slice of the earth on Eagle Peak Road. In the nineteenth century, Julian was settled by former Confederate soldiers and freed enslaved people.

In 1869, Fred Coleman, a formerly enslaved Black man married to a Kumeyaay woman, was watering his horse in another creek and saw flashes of light reflected from the bottom of what is now called Coleman Creek. Hidden today in brush and trees that have grown since then, the creek runs through town and behind the Purple Owl, a coffeehouse where, right now, I am writing these sentences. Coleman's discovery launched the second California gold rush. Miners came to cut the shafts and tunnels that weave under this land like termite tubes. Waves of prospectors pushed and pulled silt from the pools and streams in their sluices and pans, picking out flecks and nuggets of gold.

You can still find tunnels, or the mouths of them, in these mountains. I found one myself, and crawled through it until the light was gone.

The miners eventually disappeared or stayed on to become ranchers or farmers and then builders and bikers. Today, families bring their children here to experience snow, some for the first time. Their parents want for them what they had before they moved to Southern California. On the way, they pile up in traffic jams for miles. Desperate for the nature they once knew, they bring sleds for their children or lift local garbage can lids to use as snow saucers. And the town and hills fill with the shouts and screams of children.

All these voices go into the earth.

And now and then the voices surface, along with beer cans and the bones of giant sloths and mastodons and a lost cow.

CHAPTER 21

Grandfather Hunting

Most people who love nature go through a seldom-acknowledged metamorphosis over time. For some of us, age has a way of moderating consumption, calming us, setting goals that once might have seemed low bars but now feel like graduation.

Jon Young, who lives in New Mexico and teaches bird language, was introduced to nature primarily through fishing. He still fishes, but he's ambivalent about it. He says he just can't give up the stirring that happens deep in his chest when he tracks or fishes.

"I get the problem, the contradiction," he told me one day, when we were talking about what he called the tug. "But personally, I would rather catch and release than not catch at all. Fishing has tugged so many of us into deep nature." When he spent time with Indigenous hunters and gatherers in South Africa, he learned about the tug as something quite specific. "They call it the tugging. The tug pulls from down here." He points to a spot equidistant from his navel and his sternum, the solar plexus, a part of human anatomy rich in nerves. "They point to that tugging place as a point of communication. They use it when they're hunting or tracking. They say it's a primary sense, a communication with nature that's necessary for life."

He tells a story of tracking a gray fox when he was a boy. He had studied its patterns for weeks and learned its pathways, its usual tunnel through the bushes, until one day he decided he wanted to be close to it, to touch it. Just touch it, as if counting coup — the term for the act, still practiced by some trackers, of touching a wild animal without hurting it, as a show of bravery. Young's goal was to touch the fox as it passed through the woods near where he grew up.

One early morning, he stationed himself along a pathway that the fox often traveled at that specific time. He covered his scent and skin with mud and leaves. He crouched in the bushes, holding still. He said he felt something coming before he saw or heard it. Maybe he picked up the signal from the pattern of bird language around him in the woods, or maybe it was a tip-off from that tug.

The animal approaching was not a fox but a white-tailed deer. Young slowly thrust his fingers through the leaves and waited. He held his breath, shut his eyes. And finally, he felt the fur moving through his fingers.

THESE DAYS, WHEN MY SON Matthew, now thirty-six, visits and we take long walks, he approaches a particular fence and leans against it. A bull will approach and cock its head and come so close that Matthew reaches out and strokes the top of its snout. Matthew says cattle are his favorite animal. A month ago, researchers documented the first cow to use a tool; in this case, a broom to scratch herself.

Matthew is a strict vegan. And yet he still loves to fish. So do I, though the urge is usually reserved for when my sons are available company. We take great care with the fish, using barbless hooks, playing the fish quickly to limit the buildup of lactic acid, which can kill the fish later, then releasing it carefully. How strange that after hooking and yanking the fish out of the water, we bend over, cradle the fish in our hands, and move it back and forth in the water to move oxygen through its gills. Off the water, we remind ourselves that no one monitors the health of a trout stream or lobbies for stocking native more than fly fishers.

We know we're rationalizing. We still fish. Until, someday, maybe we won't. Lately, my son has been talking about using flies with no hook at all, just to watch the rise of the fish.

My son is a vegan in part because of what our carnivorous habits and all those feedlots do to the climate but mainly because of his respect for other life. A few years ago, after a day on the water and as we were reloading our belly boats (float tubes designed for fishing) into the back of our Honda Element, Matthew stopped and peered at an aquatic flying insect on the back bumper. It was probably one that he imitates when he ties flies.

He reached out and let the insect climb into his hand, then cupped his other hand over it, walked down to the water's edge, and lowered it to the water. It skittered away on the surface. As he walked back to the car, I asked, "Why did you do that?"

"Because I'm a vegan, Dad."

The word *Dad* was filled with that familiar *Isn't it obvious?* tone. Part of him is still a teenager. "Well, it's also because you take after your mother," I said, smiling.

Kathy carries spiders out of the house to let them go, and so do I (usually), partly because Kathy sets the example in our home. Plus, I've learned about some of the complexities in the lives of spiders. And eating turkey for Thanksgiving? Forget it. They're neighbors. What about chicken? It's complicated.

NATURALIST, EDUCATOR, WRITER, and former Marine Steve Ramirez is the author of five philosophical books about fishing. To research *Casting Homeward,* he traveled across the continent to spend time on some of the most legendary fly-fishing rivers — exploring both the fishing and the protection of these waters. He struggles some with the paradox of fishing and loving nature.

If catching fish were his only reason to be a fly angler, he probably would have stopped fishing long ago.

"When walking toward a stream," he told me in an email, "I practice mindful awareness. I notice the temperature of the air, and later, that of the water, because it impacts the behavior of the fish, and me. I notice cloud cover, wind speed, and stream flow, for the same reasons." With each step, he learns new things about the conditions of the trout's world. He watches grasshoppers leap and crickets creep among the native grasses "while silkworm caterpillars descend like special forces troops from the trees as dragonflies, damselflies, and sulfur-colored mayflies dance over the water."

He arrives at the stream and touches the water, feels its coolness, and then flips a half-submerged stone and inspects it for mayfly larvae, caddis fly cocoons, and the empty husks of stone flies. "I scan the stream bed for crayfish and sculpins while noticing the herons and kingfishers that tell me this stream is alive with minnows to feed the trout. Ospreys and eagles search for the same fish that I do...In nature, everything is eating and being eaten; everything is being born, reproducing, and then upon death rejoining the real circle of life. Even which wildflowers are blooming gives me hints to what the fish are doing in the here and now." And then he waits.

"At first, I sit by the stream, noticing where the trout might be in the eddy of a rock or fallen tree, or the undercut grass-covered embankment of a river bend, where the current is easy but still quick enough to carry food downstream," he says.

He imagines himself as a trout facing upstream as he watches for mayfly duns or wayward grasshoppers floating toward his feeding lane, while keeping an eye on the shadows. "And as I enter the trout's world in my imagination, I think of the messages I might send from below the water to the predators above." In those moments, he feels vulnerable to wings and talons, or fishing rods and lures made of feathers. That sense of insecurity serves to heighten Steve's awareness of everything, but particularly the possibility of mountain lions, bears, wild javelina, and feral domestic pigs. "If a predator sees a flash of silver-white, they know I am feeding along the bottom on nymphs, crayfish, or sculpins. If they see a swirl or bulge, they will know that I am feeding just below the surface on emerging mayfly nymphs."

He treats every fish with respect and care, holds the fish gently in his wet hands as he slips the barbless hook from its lip. "As a human being with asthma, I understand how important it is to keep this beautiful aquatic being in the water so that it can breathe. I release him back into his cold-water home as quickly as possible. And when I do, I say thank you, and I mean it. After all, by entering his world, I am once again reminded of the alchemy and impermanence of the fish's life, and mine. This is a lesson worth repeating."

He acknowledges that opinions on fishing ethics will vary.

NOT LONG AGO, I visited Grandfather Mountain, a nearly mile-high peak in a North Carolina state park. "This is my mountain" is the park's catchy slogan. Many visitors find themselves repeating it. After a while, I said it, too. At the nature center, I talked with a young wildlife biologist. He said he grew up fishing but mainly hunting with his father. Each decade, his father became less inclined to pull the trigger. At first, the father focused on helping his children learn to hunt for both food and sport. Now in his seventh decade, the older man still goes into the woods, sits in the deer blind — a platform high in a tree — and waits for deer to step out of the forest. He doesn't shoot, though. He just waits, watches, and listens to the forest sounds, to the birds talking in the dark.

He calls it *grandfather hunting.*

Anglers, hunters, trackers, and farmers aren't the only noticers concerned with the ethics of connecting to the natural world. One of the most benign methods is wildcrafting, the previously mentioned ancient practice of searching for and harvesting uncultivated plants usually in wild habitat for food, medicine, or crafts. The Volcan Mountain Foundation, committed to bringing people to the preserve as well as preserving the mountain, offers classes that teach people how and what to seek and take from the forest and other ethical best practices. Among them are protecting endangered species, avoiding resource depletion, making a commitment to developing your knowledge and connection to the natural landscape, deepening a local sense of place, and supporting long-term regeneration of plant life.

Gardeners debate what constitutes a native plant (a more complicated question than it may seem), how or whether to use fertilizer, and other thorny issues. Such discussions will become more complex over time, now that science is showing how much communication — and perhaps consciousness — plants have.

In ancient times and even today, some farmers and ranchers relied on their observation of celestial movements, including lunar phases, as well as cycles of birth and death, changes in animal behavior, plant growth patterns, weather shifts, and cloud formations to make decisions affecting their livelihoods. For example, cloud watchers understand that towering cumulonimbus clouds with flat bases and anvil-shaped tops are likely to bring heavy rain and hail. Wispy clouds at high altitudes indicate an approaching warm front. Clouds that create halos around the sun or moon warn that rain or snow will occur within twelve to twenty-four hours. And patches of white or greenish clouds patterned like fish scales forecast thunderstorms later that day.

Animals tell stories about the future, in part because they sense changing atmospheric pressure before humans do. Look for birds flying lower than usual as they seek shelter from a coming storm, ants building higher mounds in preparation for wet conditions, mosquitoes swarming before a storm and then disappearing an hour before the storm arrives, and cows lying down in a field before impending rain. Certain flowers, sensing changes in humidity, close their petals before storms. The leaves of oaks and other trees may turn upside down as humidity increases and rain approaches. Even telling time is possible with natural indicators instead of watches and calendars on the wall or screen of your smartphone — time as told by bird migrations, flowering plants, and so many other ways to notice.

If Indigenous populations and their knowledge and practices are further diminished or destroyed, many of these skills — more than we know — will fade, leaving us increasingly dependent on technological tools and systems, which are vulnerable to political whim. The same is true of the deep traditional knowledge of farmers. In 1980, the US Census Bureau stopped counting farm families because there were so few of them.

Concern about the disappearance of small farms and ranches is less about nostalgia than about a collapsing bank of knowledge unfamiliar to the rest of the culture, including many conservationists and environmentalists. Independent ranchers and farmers are physically closer to the land; they have dirt under their nails. That means something.

Meanwhile, big agribusiness reduces nature to spreadsheets and leaves it dependent on a vulnerable energy grid and a functioning internet, each as unpredictable as the weather.

Making peace with nature depends on another question: Can we make peace with each other?

CHAPTER 22

The Soldier and the Egret

T HE WOUNDED EARTH needs radical care. So do we.

Every few days, I go for a walk with Mark Strom, a new friend. At seventy-seven, he's still a young and remarkably strong man, despite a recent hip replacement and shoulder surgery. He retired years ago as the director of roads and parks for the city of Escondido, north of San Diego. He is married to Mary Jo Strom, one of the most effusively charming people I've met. They live down the hill from us, across the road from the acorn-grinding rock.

From their deck, they see the same Volcan Mountain view that Kathy and I see from our house.

Mark grew up on the beach in the sun north of here. As a kid, he was painfully shy, and like many young people, he discovered that other animals were more comfortable to be with than humans. He often brought a damaged animal home to care for until it could be released. When he was fifteen, he rescued an injured crow, nursed it back to health, and named it Joe.

The crow learned to imitate the crying baby who lived next door. And the sounds of the neighborhood cat fights, too. "*You* don't have to listen to him all day," Mark's mother pointed out, referring to the crow's calls. She was usually very patient with Mark and his animals. But on that occasion, she was not. When Mark got home from school one day, she told him he needed to set Joe free. Now.

As a young man, he married, fathered a child, and enrolled in college. In 1969, when he failed to enroll in enough college classes to keep his student deferment, Mark was drafted.

One symptom of post-traumatic stress disorder (PTSD) is the need to relive the trauma again and again. Mark often recounts the cruelty he witnessed on both sides — including the acts committed by some of his fellow soldiers — the violence among them, his own role as a machine gunner, the horror of carrying men, often missing their limbs, to the helicopter.

On a recent summer evening, I was sitting with Mark and Mary Jo and their dog Lillie on their back deck, watching the fog move slowly toward Volcan Mountain. We were thousands of miles from Vietnam, but Mark took us there. He told a story about a tiger[vii] who one night crept out of the forest, bit into a sentry's boot heel, and dragged him toward the trees. The soldiers intervened and yanked the sentry to safety.

Even when immersed in the horrors of war, Mark felt released by the natural world. Lifted. Mark waded into a creek to help some young Vietnamese boys fish. They were about to turn draft age. Suddenly, what looked like flying fish exploded into the air and seemed to vanish. They were mudskippers, amphibious fish that walk on land and breathe air. "They look like a cross between big tadpoles and tiny birds," he said. They landed on a rock. Sitting perfectly still, almost meditatively, they watched Mark and the boys.

His fellow soldiers gave him two nicknames: Muscle Beach, because he was a bodybuilder and surfer, and Nature Boy, because he protected the wildlife they encountered.

While on patrol, he and his fellow soldiers encountered a white egret sitting on an open patch of ground. It appeared to be injured. Mark picked the bird up and held her under his arm. She didn't fight him. "We moved through an opening in the trees, and these guys go, 'Wait, Muscle Beach. Man, what are you going to do with that *bird?*'"

---

vii   The last known sighting of a wild tiger in Vietnam was in 1999.

They thought its white feathers would draw fire.

"I said, 'No, it won't. Egrets are here all the time.'" Which wasn't entirely accurate, because so much of the wildlife had been killed or scattered by the war.

For two days and two nights, Nature Boy carried the egret under one arm and his M16 under the other. "We would go to the highest hill and dig foxholes, put out our Claymore mines and our trip flares. I'd settle the bird down next to my rucksack." She would sleep with her head and long neck tucked under her body. Sometimes the men and the bird slept in shallow water. To feed the bird, Mark opened cans of chopped ham and eggs. "We all hated that stuff. The bird loved it.

"Then there's a firefight. These guys go, 'You got to get *rid* of that f—ing bird. Kill it! Or *we're* gonna kill it. Or you.'"

On the second day, at dawn, the egret stood straight up in the foxhole and extended her neck, stretched her wings, looked at Mark and lifted into the air. She flew into the air, then over the tops of the trees, and disappeared. At that moment, Mark felt an intense sense of *lift.* He held that feeling tightly for the rest of the war and has held it ever since.

A few weeks after the egret flew, Mark was wounded during a firefight. The army shipped him to the United States for medical care. When he recovered, he was sent back to Vietnam. There, he learned that most of the soldiers in his platoon had been killed or wounded.

A FEW YEARS AGO, after raising his daughter, Mark moved to these mountains. There is something about our local mountains and the desert below — perhaps the isolation — that draws war veterans here to heal, if they can.

Every war is a war on all life. Walking the narrow roads on the west end of the valley late one afternoon, I watched as an invading ant army, a species that has migrated here from Central America, attacked and devoured smaller native black ants. The black ants are preferred — even required — by horned lizards. The disappearance of native ants is one reason horned lizards have grown scarce.

Most summer evenings, Mark and Mary Jo sit on their back deck. As he holds a glass of wine, Mark will watch the horizon. And wait.

Tonight I've joined them. In the fading light, Mary Jo goes inside to make dinner. Lillie and an orange stray cat named Pumpkin sit with us. When we are about to call it a night, a ghost materializes at the edge of darkness. Far away, it dips in a downdraft, rises on warm air, comes closer, and then glides across the valley, as it does on most summer nights.

We watch the white egret disappear beyond a ridge, like a light going out.

We're both smiling. I wonder aloud why all these animals have come into his life. Mark answers, "To bring me peace."

VI. OUR HUMAN-NATURE SPIRIT

Life is a garden, not a road. We enter and exit through the same gate. Wandering, where we go matters less than what we notice.

— KURT VONNEGUT

We respect and take care of this land because she has always taken care of us. She is our mother.

— KUMEYAAY ELDER DR. STANLEY RODRIGUEZ

CHAPTER 23

"Are We Lucky?"

Around the bend from us, I often stop to talk to my friend and neighbor, Jim Baker, a retired fire chief of Del Mar, California. A long time ago, his fellow firefighters nicknamed him Bake.

Usually, when I see him, he grins and points to the trees and the sky and says, "Are we *lucky*?"

Jim is ninety years old. He lives alone in a brick house he built himself. Between the house and the road is a long brick wall. He built the wall as a gift for his wife. A pedestal stands at one end with a plaque that says Wall of Hope.

One day, Jim looked out his front window and saw a middle-aged woman he did not know sitting on the wall. A few weeks later, he saw her again. And then again. Finally, he walked down and talked with her. The woman told him she drives nearly an hour to sit on the wall. He asked her why.

"Because it gives me hope," she said. She visited the wall for a year.

Next to Bake's house is a little grave plot where his dogs are buried. Two of his four daughters, laid to rest elsewhere, are also commemorated in the plot with markers. For many years, Bake's wife suffered from Lewy body dementia. After she died, his former firehouse colleagues, who love him, carried her in a firetruck to her burial in San Diego. Afterward, Jim placed a porcelain flower next to those markers.

And *still*, he feels lucky every day.

"We live in Lucky Town," he'll say. "We'll make it." Then he adds, "Hope is only the first part."

Fire has ravaged these mountains multiple times in the past twenty-one years, and the fires are getting bigger. In 2003, a lost hunter started a signal fire about nine miles southwest of Julian. The fire grew overnight and roared into San Diego.

It burned large swaths of the suburbs and came within a few blocks of our home in the Scripps Ranch neighborhood. As the fire neared, Kathy, Matthew, and I scrambled to evacuate. Jason was away at college. Binkley, our cat at the time, was missing. Then we heard him under our bed. As I reached for him, Binkley pulled back. Matthew began to cry.

"Hey, Matthew, don't worry. Could be worse. I grew up with tornadoes," I said, stupidly.

"Well, I *didn't*."

We packed up, with the cat, and headed for the coast. We were lucky.

Growing up, I often heard and read about the triumph of the human spirit. The term was usually associated with the horrors of the Great Depression and World War II. The human spirit represented people's ability to rise to a great challenge, particularly when so many people were out of luck.

Like many people, I push back against a tempting sense of despair sitting like a shadow in the corner of the room. But it isn't easy.

To deal with despair, Anthony Paul Smith, a professor of religion and theology at LaSalle University, suggests reflecting on "melancholy to see if it is, in relation to ecology, perhaps a creative emotion." He adds, "A drive to create a future can arise out of dissatisfaction with this present, with this world."[159]

However we reach *that* future, it should be informed by the natural world, by its systems of communication that stretch over vast distances within and across so many species, by its strange and permeable consciousness, by its symphony, its sky.

The best path beyond despair is also through imaginative hope — envisioning a world we want to go to, then living in its creation. Focusing only on the *human* spirit places too much emphasis on one species. Human self-centeredness is why we're in this environmental mess. We need a new spirit and a new term — one that includes all the life around us, both human and more than human. A human-nature spirit.

DURING THE 2003 FIRE, the wind reversed, as it often does in the afternoon. Gaining speed and force, it created its own weather system.

Bake stood at the overlook near his home. He watched as the sun disappeared into a wall of smoke. The twenty-mile-long line of flames was headed toward Julian. Government agencies and fire districts, each with its own specialized communication systems, remained in their silos. Some tanker planes and helicopters, including those under military control and others waiting for approval from the governor, never took off.

This was, until then, the largest mobilization of firefighting resources in California history. What became known as the Cedar Fire burned 273,246 acres, destroying ranches, fields, and forests. It killed twenty people, including one firefighter, and countless animals, wild and domestic.

Spreading at 3,600 acres per hour, the fire burned so fiercely that it shattered boulders the size of houses and sterilized the soil and seeds needed for future forests. Residents died at home or in their cars.

Jim hurried between his neighbors' homes, grabbing their garden hoses to fight the flames for hours, often alone. When the fire finally subsided, he returned home, walked past his Wall of Hope, stopped at the bottom of his driveway, and sank onto the concrete.

Volunteer firefighters took their shifts. Two days later, he was back at it again.

After the firestorm passed, he contacted some of his human neighbors and asked for their permission to enter their houses, if they still stood.

The power was still out. Food would soon spoil in refrigerators. Jim asked his neighbors if he could take their food to share with the few people who hadn't fled, as well as with the other-than-human neighbors: deer, dogs, cats, skunks, and the waddling opossums now covered in black ash.

They said yes. Some residents, in their haste to escape, had left their homes unlocked. Others told him where they had hidden their house key or where a window might still be open or unlocked. They trusted him.

Over the next few days, he wandered the roads and trails on the hill. He left chicken, stale casseroles, and bowls of milk in the lots, in the fields, or behind the retaining walls where he knew the animals gathered or where he could see them retreating into the shadows. He understood that they were singed or burned, terrified, hungry, and thirsty and that most of them had become homeless.

When he ran out of food for them, he started heading back to his brick house.

As he approached the Wall of Hope, Jim heard voices calling to him from porches and cars. People were pointing behind him. He turned and saw a parade of animals, wild and domestic, following him. Hoping.

SOMETIMES I THINK of Jim Baker and James Hubbell as the Two Jims. Both wise, both inspiring, for related but still different reasons.

Hubbell died in 2024 at the age of ninety-two.

He lived in these mountains for decades. During that time, as an artist and a human being, he taught thousands of people around the world — usually indirectly — about trusting life.

After a childhood of isolation and rootlessness, and too many men who passed through his mother's home, he came here as a young man and chose to live along the watershed below Volcan Mountain with his wife, Anne. Here, he made his own, new roots.

Their first adobe building held only one small windowless room. Over time, they made a heart-filled home and a compound of art studios from the earth. Each building is a work of art. From then until the present, students have come from around the world to learn from James, from nature, and from one another. He and his students also designed and built multiple symbolic gateway sculptures in countries around the Pacific Rim — gateways to art and nature, to immigrants, and to the faith beyond hope.

In his journal, James once wrote, "The interconnection with all life is not a random thought. To me, it is a fact. To say that we are more alike than different is true, as human life was formed from stardust, the explosion of stars to form supernovas that create and cast all the elements of the planets and all life in the universe. It gives me comfort to know that life has this continuity. We are born containing these starburst elements, and these particles and energy remain, even after death."[160]

"Trust life," he liked to say. It was his motto. Trust life.

I wasn't as satisfied with that motto as he was. "But don't you worry about the next fire?" I asked.

He answered, "It's your gift to worry."

Since James's death, I often reflect on his views about trusting life. He frequently talked about life's ability to regenerate and evolve, regardless of circumstances. He also emphasized authenticity. Nature is authentic. James Hubbell, like Jim Baker, is authentic — they're the real deals. So, I believe the reason to trust life is that life is true. While life can end, it cannot be manufactured — at least not yet. Recently, on a whim, I asked Google, which uses generative artificial intelligence, this question: "Is life true?" Its answer:

"Yes, life, in the sense of our conscious experience of existence, is real. While the question of whether our current reality is a simulation or not is a philosophical debate, the fact that we are thinking, breathing, and interacting with the world around us points to the reality of our subjective experience."

We? Really?

When AI begins to refer to you and me and itself as "we," we might want to take a break. So much of life feels like a description of breath rather than the breath itself.

In daily life, in politics, and in internet searches, authenticity seems increasingly in short supply. The wind of technological experience is gradually taking us beyond the tactile, ever further from what our five or thirty or fifty senses can detect, moving us every day toward the unreal — or the unalive. Which is at least part of the explanation for why we (or at least we humans) hunger for what we call nature: the look in a fox's eyes as it turns toward us in the forest with its singing trees. Moments like these are as close as we can get to the larger force that moves us, the *life* force to which we belong.

I later asked Google an additional question: "Should we trust life?"

"An AI Overview is not available for this search," it answered.

JAMES HUBBELL AND I stood one day at the edge of his property, and he shared a vision of what he believed would happen to him someday.

He said his ashes would be buried in the dirt on the side of the mountain, and when the rain came, the ashes would flow down into a creek, into a river, and into the Pacific Ocean — and then travel on an ocean current to a distant shore, where what remained of him would give life to a seed. A flower would open from that seed. And a young girl on a walk would see the flower, pick it, hold it up to the light, and admire its beauty.

The metamorphosis. The hope beyond hope.

At that time, James was battling late-stage Parkinson's disease. He struggled to control his limbs. "Trust life," he still said.

But one day he surprised me. We were talking about the future. He said he does not believe the human species is sustainable. "But trust life," he said again.

His work remains, surrounded by the new life that always emerges from the ashes of previous fires, much of it visible on the Hubbell hill, in those strange and moving buildings and sculptures that came from his life. A nonprofit now oversees all of this.

When our mutual friend Peter Jensen first met James Hubbell, he expected the artist to give him a tour of his compound.

Instead, James said, "I smell smoke. Let's go see where it's coming from."

They set out on a path that snaked between and over the hills of grass and trees. On one hill, they could see a cluster of sentinel oaks, and near these trees was a single hollowed-out oak with bare and fallen branches. Inside the trunk, they could see an ember the size of a small child, throbbing and glowing red.

Peter's first response was to feel sad for the once-majestic tree.

Then he looked at James.

The artist, who believed that beauty will save the world, was grinning. He pointed inside the oak and said softly, "See? Its heart still beats."

CHAPTER 24

The Animate Circle

Most of the time, people assume ownership of where they move. But scrape the surface of where you live or where you pass by, and possession becomes an abstraction, as fleeting as the clouds that follow their shadows across the land.

Mac Macartney wrote *The Children's Fire,* a book about the early tribes of what became known as the British Isles. He once told me he came to the United States to learn from Indigenous elders. When they asked him who he was, he answered in Western terms, offering his name and occupation. They replied, "No, who *are* you? What tribe are you from?" Macartney, a Welshman, told them he had no tribe. "Yes, you do," they said. "Go back and find your tribe, and you'll know who you are." He did just that, discovering his roots in the ancient Celtic Briton tribes, on his mother's side.

Few of us, regardless of bloodline, remain in our First Place. We move on and hunger for what we barely know we missed. Which is a reason to consciously attempt to notice and know the land we are dealt or choose, so it can at least take root in our hearts. And just because we can't locate our original tribe doesn't mean it's not important to our roots.

Sylvie Rokab, a mindfulness teacher and filmmaker (notably of *Love Thy Nature,* narrated by the actor Liam Neeson), recently contacted me.

She wrote, "Coming from a family of refugees, I understand the challenge of deepening connection with place when 'home' is a moving target. What has helped me, as well as other people that I work with, is what in Buddhism is called the beginner's mind, that sense of perceiving every moment, person, and situation anew." She believes that becoming more aware of impermanence as a natural part of life can bring comfort to those facing the challenges of displacement.

As the tectonic plates of climate and life shift at an increasingly rapid pace, our descendants' ability to survive will require both solitary immersion and social connections — and questions. Whether we're moving to a new place or staying put and planting deeper roots, we need to ask: What animals and plants are disappearing from our environments? Can anyone identify and understand the incoming species? What are the skills of noticing and connecting that we and our great-grandchildren will need? Which of these skills can we start practicing right now? The future will belong to the nature-smart. But to become nature-wise? That's more challenging.

WHEN WE MOVED to these mountains, I rediscovered my childhood sense of being surrounded by circles, orbits, and all those satellites of life out there.

I recall the black earth, then the corn roots. And dandelion roots, which usually descend a foot or so into the prairie soil but which, when undisturbed, can reach down fifteen feet. Beyond the corn, more circles contain the starlings, killdeer, and grasshoppers in their own domains, and then the houses viewable on the west side, beyond the next horizons to the east, north, and south. All the wildlife in circles visible and invisible, then the natural woods and, hidden behind that horizon, farms in their own circles, then the round earth, and beyond.

This awareness of life's concentric circles may have only been the focus of a child at the center of his own universe. But sensing all those circles, I wondered who lived in them and what their lives were like, and I thought about the life of the wind and the coming rain.

I climbed the oaks in the woods — elms, cottonwoods, poplars, silver maples, green ash, and spruce — all planted decades earlier to prevent the black topsoil from blowing away. I didn't understand then why those straight lines of trees were there.

Tallgrass prairie, with roots ten or more feet deep, held much of the Midwest, including this part of Missouri. Farms and ranches have destroyed much of the tallgrass prairie, but some remains, especially in the Kansas Flint Hills.

Nature made circles; people added lines.

Driving across the plains, you can see those windbreaks still. But today, many of them are disappearing. Cash-strapped farmers and factory agriculture make room for more row crops. Because of drought, climate change, and the cutting of those trees, we may yet witness the return of the dark storms that blew away the black soil in the 1930s.

The great columnist Ernie Pyle, who died during World War II, once wrote, "To me, the summer wind in the Midwest is one of the most melancholy things in all life. It comes from so far away and blows so gently and yet so relentlessly...It just keeps coming...You could — and you do — wear out your lifetime on the dusty plains with that wind...And when you are worn out and gone, the wind — still saying nothing, still so gentle and sad and timeless — is still blowing across the prairies, and will blow in the faces of the little men who follow you, forever."[161]

Melancholic or not, that wind ignited the senses and the imagination, made the leaves clatter and rise in sudden dust devils. They would whip and stretch higher, like little headless tornadoes. My friends and I would run to stand in the dust devils, and holding on to our ball caps, we'd look up through spinning wind, past the leaves, toward the circling universe.

ALL THESE YEARS LATER, I imagine a circle around our house — what I call the *animate circle*. Most people would pronounce *animate* in a way that suggests creating a moving cartoon. But the adjective *animate* — pronounced and perceived as the opposite of *inanimate* — feels closer. Within this circle are other concentric circles, where roots reach and creatures eat and sleep. These aren't real borders, but imagining them helps me keep track of where the plants are and where the animals move. It's a kind of mental map.

An animate circle both gives life and is life. When two people enter each other's circle of awareness and they bond, they begin to share each other's thoughts and feelings. Sometimes they finish each other's sentences. The process can also occur with a tree or a forest. Or a deer.

On my walks these days (though not during hunting season) I can move closer to the deer I encounter. I enter what biologists and hunters call the deers' "flight zone." They've become more accustomed to me. But take one step deeper, cross a line invisible to me, and the deer fling themselves into the air and bound away, stopping only when they reach the next animate circle.

The Canadian scientist Suzanne Simard, best known for her pioneering research on the underground networks of fungi in forests, has said, "Our aboriginal people...don't even have a word for 'the environment,' because they view trees and plants and animals, the natural world, as people equal to themselves. So, there are the Tree People, the Plant People; and they had Mother Trees and Grandfather Trees, and the Strawberry Sister and the Cedar Sister. And they treated them with respect, with reverence."[162]

Today, nature therapists attempt to tap into these older cultural roots, sometimes successfully.

THE CIRCLE — the big one and all its concentric offspring — has returned to my daily awareness. Each place is at the center of its own animate circle. Circles create and contain our sense of home. A circle of influence can exist in a yard or a wild urban park, can contain several blocks or square miles of a city, a forest, or farm country, or, as our awareness and care expand, can even encompass a whole bioregion and beyond.

Within this animate circle, we can feel physically, psychologically, and spiritually connected to other lives and our own existence. Focusing our attention on kinship, we can get to know the wildlife, trees, vegetation, rock formations, and weather patterns as well as the tunnels and rock castles of ground squirrels and the haunts of the fox. We can also gain a clearer understanding of the deliberate and accidental impacts of our presence.

Native or edible plants are included in our animate circle map, along with humans. Where the miner's lettuce grows and the forest ends. Within the circle, we mentally locate Jim and Mark and Mary Jo and our other friends on the hill, as well as the phone line repair person, and the people we chat with about the weather or that fox.

Trail cams offer a way to explore the circle and share our discoveries. On our hill, an online circle of neighbors does this regularly by attaching photos and videos to texts or posting them on the neighborhood Facebook page. This neighborhood tradition of sharing trail-cam photos and wild-life videos began during the pandemic and continues today.

The animate circle is also animated by time.

My friend Mike Houck founded Portland's Urban Naturalist Program and works through the Intertwine Alliance to create wildlife corridors and parks in the Portland-Vancouver area. As a result of his efforts, beaver-friendly culverts were designed and built. Portland now has a healthy beaver population.

He told me this story: Walking daily through a park to work or school, Mike noticed the same bird every morning. He would talk to the towhee as he passed. He did this for years. One day, he realized that the ritual had continued well beyond a bird's natural lifespan and that he had developed a relationship with generations of the same species. In a sense, he had bonded with a bird that never died.

CHAPTER 25

Healing the Ecological Self

WHILE STRANDED at the Phoenix airport one afternoon, my friend Kathleen Lockyer, a noted ecotherapist and nature educator, shared a story with me. We were both heading to the same conference and would miss most of it. Her story, which she later included in her book *Wild Inside*, brightened the day.[163] Here's the story, in short form:

One day, Kathleen was surprised to see her often-hyperactive seven-year-old daughter kneeling on the ground and staring over the edge of a rusty water trough, perfectly still.

She quietly joined her daughter and looked over the same edge. The bucket was half full of water, but that was not the target of her daughter's gaze. A dragonfly nymph emerged from the water, "dragging its tiny water-logged body eight inches up from the edge."

There, it began the slow and laborious process of shedding its muted gray shell. She watched her daughter's body remain still, the child's eyes wide, her hair tangling in the wind.

They observed the shedding for an hour. As Kathleen recalled, it was as if her daughter were watching herself emerge.

The new dragonfly finally reached the top edge of the trough and perched, wings spread, moving them slowly as they dried. Kathleen said to her daughter, "Wow! No mother. No father. How does it know what to do?"

Her daughter softly answered, "Mama, maybe, just maybe, she dreams a dream her whole life until one day she wakes up and just knows what to do."

Kathleen takes great pleasure in telling this story.

She saves the ending for later.

EARTH'S ENVIRONMENTAL CRISIS is measurable not only in the number of fires, floods or inches of rising seawater but also in the rising rates of depression and anxiety, some of it attributable to what is now called eco-anxiety or eco-fear.

The American Psychiatric Association defines *eco-anxiety* as "a chronic fear of environmental doom." The *Lancet,* Britain's prestigious medical journal, reports, "Symptoms associated with climate anxiety include panic attacks, insomnia, and obsessive thinking…potentially leading to increases in stress-related problems such as substance use disorders."[164]

Eco-anxiety has long-term implications. In 2016, a report issued by the American Psychological Association, Climate for Health, and ecoAmerica (a nonprofit for which I served as a charter board member) projected long-term societal damage caused by eco-anxiety. Among other problems, the groups cited interpersonal and intergroup aggression and loss of social identity and cohesion, especially for "Indigenous communities, children, and communities dependent on the natural environment."[165] Research now suggests that mental health professionals who treat eco-anxiety are increasingly seeking personal counseling for their own environmental depression.

One response could be to strengthen our shared ecological self. The concept of the ecological self originates from deep ecology, an environmental philosophy developed by Norwegian philosopher Arne Naess in 1972. Naess argued that human life is one of many elements in an ecosystem. The ecological self is an expanded sense of self, one that includes all of life. In this view, we are not the entitled apex of life but are part of a larger identity that encompasses the rest of nature. We are more than ourselves.

Today, the public health approach called One Health maintains that human illnesses cannot be treated only by treating a human body; the health of all life in, say, a city must also be attended to. By destroying the environment, we destroy ourselves. By reviving the natural world, we revive ourselves.

We see a rapidly expanding body of scientific knowledge about the links between nature and physical, cognitive, social, and psychological health. The number of studies has grown from approximately sixty around the turn of the new century to over fifteen hundred today.

This new body of research also suggests that increased exposure to nature reduced the symptoms of attention deficit hyperactivity disorder and prevented or mitigated obesity, myopia, and vitamin D deficiency. Premature births and other critical health measures are better in neighborhoods with more parks and trees and worse in neighborhoods lacking these natural features. The benefits of nature connections may be universal, but access to nature is not. For many, the connection is a matter of life and death. In addition, integration of nature experiences into education is linked to more creativity, improved executive skills among children, lower burnout rates among teachers, and even higher standardized test scores, among other benefits. (Summaries and links to this research can be found at the Children & Nature Network's Research Library.)[viii]

Many healthcare providers and health organizations are responding to this growing body of evidence. In the United States, Canada, and several other countries, pediatricians are now prescribing or recommending time in nature for the families they see. They view nature connections as one of the few healthcare approaches that can be both preventive and curative. Both the American Academy of Pediatrics and the American Medical Association have issued statements recommending more time in nature.

The shape of schools and their surroundings are becoming greener. A movement for more natural school grounds is spreading in the United States and around the world. We also see the dramatically growing popularity of nature-based and forest schools, especially for children in their early years.

---

viii   The Children & Nature Network's Research Library can be found online at https://research.childrenandnature.org/

The American Psychiatric Association is also paying attention; its journal recently cited one study that found that "everyday encounters with birds are associated with mental well-being in people with and without mental illness, and the benefits persist beyond just the encounter." Birds are beneficial because they're widely abundant, even in highly urban neighborhoods.[166]

Also, the journal *Frontiers in Public Health* suggests that targeted ecotherapy can especially help with psychosocial conditions like depression, chronic stress, and PTSD. The researchers suggest that some of these benefits might primarily come from *social* contact with other species. When adults and children spend more time in contact with wildlife and pets, they often feel less alone in the world.

In her popular and influential philosophical blog, the *Marginalian,* Maria Popova has described 2021 as the most challenging year of her life, especially because she was dealing with the onslaught of bad news that year, much of it related to the pandemic.

Trying to imagine a better life, Popova resolved "to face it like a tree — a resolution blind to that unfathomable future, as all resolutions and all futures tend to be, but one that made it infinitely more survivable." Humans, she pointed out, "have a long history of learning resilience from trees and fathoming our own nature through theirs."[167]

Carl Jung, often cited in ecopsychology literature, referred to the collective unconscious — that part of the psyche shared by every human. According to Jung, the collective unconscious is a repository of instincts and archetypes, as he called them, including the great mother, the wise old man, the shadow, the water, the tree of life, and others that reveal themselves in our myths and fairy tales.

Jung did not list the dragonfly as one of his archetypes. Perhaps he should have. As dragonflies struggle to emerge, they sometimes die trapped in their outgrown exoskeleton. Or, while they are pulling their own abdomen out of the shell, a gust of wind can disrupt their transformation, end their attempt to escape from their encasement, and lead to their death. While they are this vulnerable, they can be attacked by a predator.

Sometimes they survive, sometimes they don't.

*   *   *

SO, HERE'S THE HAPPY ENDING to Kathleen Lockyer's story:

After the dragonfly's wings had dried and it was ready to launch into adulthood, a gust of wind drove it back down into the water. The little girl's hands shot out, one above and the other below the newly transformed creature.

She slowly lifted the insect out of the trough and placed it in the grass. She knelt and shielded it with her body until it dried out again. Then the dragonfly, like Mark Strom's white egret, lifted into the air and flew off toward a new future.

CHAPTER 26

The Great Noticing

To MEET THE CHALLENGE of nature-deficit disorder, people are building a new nature movement. The movement is fueled by research and the dedication of parents, teachers, city planners, biophilic architects, and others in the United States and around the world.

We face four threats that have become one: the climate crises, biodiversity collapse, alienation from the rest of nature, and human loneliness.

We can't do much about any one of those without addressing the others. But for connecting children to nature, we see action. As noted earlier, pediatricians in Western countries have begun to prescribe nature time for children and their families. One reason is that families need more exercise. They also need the healing power of nature. And they need to notice and to be noticed.

Meanwhile, as described earlier, the number of green playgrounds and nature-based schools, both private and public, is increasing. The International Union for the Conservation of Nature, the world's largest network of conservation organizations, will soon launch a worldwide campaign for nature-based education. In the United States, a partnership between the Children & Nature Network and the National League of Cities, which represents some eighteen thousand mayors and other urban leaders, has launched a campaign to rethink how cities can become nature-rich for human health and biodiversity and to combat the climate crisis.

In China, Brazil, Australia, the United Kingdom, Canada, and many other countries, regional and national campaigns are emerging to counter the growing gap between children and nature.

All of this is good news, but it's not enough.

Zen master Thich Nhat Hanh made the case that the environmental challenges we face—the climate emergency, biodiversity collapse, zoonotic pandemics, and our deep species loneliness—will require a spiritual revolution. "You carry Mother Earth within you," he says. "She is not outside of you. Mother Earth is not just your environment...Not to cut the tree, not to pollute the water, that is not enough...In that insight of inter-being, it is possible to have real communication with the Earth, which is the highest form of prayer."[168]

Some folks will find this ambition too ethereal. But Indigenous peoples have been pointing us in that direction for a long time. Increasingly, scientists are turning to traditional ecological knowledge each year. The health of the natural world, and our survival in it, will depend on our species' ability to love the natural world, to become re-enchanted—bioenchanted. And to re-become.

A natural empath, Jack Kerouac seems to have had a hypersensitivity to other human beings as well as to the rest of nature. His senses were always on high alert, sometimes to his detriment. His life was short and ecstatic and sometimes despairing, but his writing reveals an exceptional empathy. As he traveled America's bedraggled backstreets or stood in a fire lookout above Big Sur, Kerouac saw all of humankind and all of nature as the same entity, deserving of our compassion. Few people think of him as a visionary. He was famously insecure and melancholic, yet often his writing was movingly hopeful. In his journal, Kerouac wrote:

> It is suddenly occurring to me that a great new change is about to take place in mankind and in the world. Don't ask me how I know this. And it's going to be very simple and true, and men will

have taken a step forward. It will be a kind of clear realization of love, and war will eventually seem unreal and even obsolete, and a lot of other things will happen. But madness will rule in high places for a long time yet. All this is going to come up from the people themselves, a great new revolution of the soul. Politics has nothing to do with this. It will be a kind of a looking around and noticing of the world, and a simultaneous abandonment of systems of pride and jealousy, in many, many, people, and it will spread around swiftly.[169]

For many people, thinking about the future conjures up images from movies like *Blade Runner, Mad Max,* or *The Road:* some postapocalyptic dystopia stripped of nature and human kindness. We seem drawn to that flame, despite our fears, but it's a dangerous fixation. Not that there aren't reasons to be discouraged, but here's the problem with despair: As Martin Luther King Jr. taught us, any movement — any culture — will fail if it cannot paint a picture of a world that people will want to go to.

This is why, more than ever, we need a new nature movement, one that includes but goes beyond the good practices of traditional environmentalism and sustainability, one that paints a compelling, inspiring portrait of a society that is better than the one we presently live in. Not just a survivable world, but a nature-rich world in which our children and grandchildren thrive.

THE NEW NATURE MOVEMENT revives old concepts in health and urban planning while also introducing new ones in light of recent research that demonstrates the profound impact of nearby nature and wilderness on our psychological and physical health, cognitive functioning, and economic and social well-being. University of Colorado professor Louise Chawla describes the basis of this movement as "the idea that as humans we cannot only make our ecological footprints as light as possible, but we can actually leave places better than when we came to them, making them places of delight."[170]

We see growth in the fields of biophilic architecture and biophilic urban design; we observe the creation of workplaces, schools, neighborhoods, and whole cities that incorporate nature; and we then witness increases in productivity, creativity, and health.

A decade ago, biophilic design and architecture — which incorporate natural elements into building and neighborhood design — were rarely mentioned in architectural and urban design journals. Now, it's hard to find such journals that do not promote biophilic approaches.

Studies have shown that workers in biophilic workspaces, along with students and faculty in schools, are not only healthier but also more productive and happier. Similarly, adding natural areas like green spaces, tree canopies, native plants, and vegetable gardens to neighborhoods — whether new or old — improves physical and mental health, reduces crime, and can even increase home resale values. Greener cities support biodiversity, which recent research indicates can serve as a buffer against future pandemics and help lower urban heat.[171]

The psychological, physical, and cognitive benefits of nature connection may be universal, but access to natural areas is not. Some progress is being made. In September 2012, the World Congress of the International Union for the Conservation of Nature (IUCN) passed a resolution declaring that children have a human right to experience the natural world, an essential ingredient if nature is to be protected from human excess — and a step toward seeking a similar declaration from the United Nations.[172] We have a long way to go. Only when nature access is viewed widely not as a nice-to-have but as a have-to-have, will the inequities be fully addressed.

Consider the city of Oakland, California, where pediatrician and clinical scientist Nooshin Razani is the founder of the Center for Nature and Health, the nation's first nature-based clinic associated with a major health provider, University of California, San Francisco, Benioff Children's Hospital. Razani says that the neighborhoods with the most poverty are often those with the fewest parks and the least green space. "This is a social justice issue," she says. More than 90 percent of her clinic's patients are on Medi-Cal, living at or below the federal poverty line. And many suffer from a lack of nature access and from pollution. According to a study published in *Environmental Health Perspectives,* African American, Hispanic, and low-income children are most likely to be exposed to toxins at school. One reason is that school districts in less affluent communities often place schools on cheaper land next to highways, factories, or contaminated sites.[173]

Adjusting urban access to nature will be a monumental task. UK researcher Miles Richardson, whose 2025 study found a 60 percent reduction in human-nature contact over two centuries, told *The Guardian* that "a city may need to be 10 times greener to reverse declines in nature connection" and that even with dramatic urban greening and enhanced nature engagement, "projections suggest a persistent disconnection from nature through to 2050." But here's the good news. After 2050, transformative efforts to bring more natural habitat into cities "could trigger a self-sustaining recovery."[174]

To slow climate change and reverse biodiversity collapse, Ross Andersen, senior editor of *The Atlantic,* suggests that "we need to reimagine our entire relationship to the nonhuman world." And with ourselves. "Maybe," he adds, "it's one of these deep problems that, if we engage it in a serious way, changes us. Maybe we need to become different people in relation to the natural world. And maybe that isn't such a wildly utopian thought: that becoming different people is something that humans do, in wrestling with deep problems."[175]

One question that emerges is how an aptitude for noticing and caring about nature can be passed to our children and grandchildren and to adults who skipped that special stage of immersion in nature during their childhood development. Until recent decades, countless generations have experienced this stage, which has shaped their spirits. How do we, as parents, teachers, and mentors, revive that spirit in ourselves and pass it on to future generations?

We can continue to teach traditional science but, at the same time, expose more students to the science and spirit of bioenchantment.

* * *

UNTIL SHE MOVED TO JULIAN, Patti Shields operated a nature-based school in San Diego. One morning, I ran into her at a town coffee shop. We sat outside in the sun and talked about imagination. I asked her how the use of imagination in nature related to her teaching.

"I was that teacher who always had critters in my room," she said. "I had one child who only spoke when he was holding a guinea pig." Each guinea pig went home with a different child for the weekend or holiday break. The students would care for them there, then bring them back to school on Monday. Shields would ask the kids to write stories about the guinea pigs and other critters (mostly rescue animals). Her students often illustrated their stories with photos or drawings. She could tell that the children had caught a glimpse of the animal's essence. "These activities required imagination — suspending so-called reality to allow my students to become something else, maybe even go somewhere else. We couldn't exist without this deep level of imagining, leaving our current existence to make space for new ideas and wonder."

For imaginative hope, we only need to ask, What if? And then, Why not?

What if educators continue to teach traditional science but also expose more students to the science and spirit of bioenchantment?

What if more students — and adults in intergenerational programs — could learn about deep noticing and their supersenses, those thirty or so doorways of perception? With their teachers and a few visiting parents, they might use critical anthropomorphism to study a snake, review the complex science of how a snake's tongue works and then imagine what the snake is sensing and tasting.

What if they learned the languages of other species, the networks of knowing, from spider webs to the mycelium that connects and nourishes the forest? Although many art teachers introduce students to the natural world, what if that approach were expanded, perhaps by incorporating nature journaling? On their sit spots in the woods, the students could welcome frogs back to the water's surface and feel what it might be like to live as a frog, or watch a fox cross through the high grass and wonder about its stories and its life.

What if, through the art of deep listening, they could hear the music of life floating in from the trees at the end of the playground, an urban park, or a remaining pocket of wilderness? All those voices. The animal orchestra.

Why not expand sensory gardens, often created in schools for children with special needs, to reach more students? Or create earth gardens that utilize forests, deserts, and seashores? By helping students learn to pay closer attention to nature, an earth garden could improve their cognitive skills and help young people better understand their peers and their neighbors — human and more-than-human. And why not teach about community by encouraging each child to notice and then map the animate circle that surrounds and includes their home and neighborhood?

What if reciprocity were a central tenet: appreciating the gifts of the natural world and returning them in kind. Students might steep themselves in how wildlife and plants perceive place and time. Through the study of geology, they would learn the stories of the earth. They could learn regenerative agriculture, biophilic design, and how to increase urban biodiversity, beginning in their yards or on their rooftops. They could learn about nature therapy and ecopsychology and how to incorporate nature connection into every kind of career. Such an education, much of it beyond school walls, would contribute to the ecological health of cities and towns and to the regeneration of democracy.

Through the pursuit of deep noticing, students and teachers could practice imaginative hope and, through this process, begin to reimagine cities, neighborhoods, workplaces, schools, and parks as places of health for the children of all species.

IMAGES AND WORKING MODELS of a future worth building are forming now. They incubate in universities, tinkerers' garages, corporate boardrooms, everyday people's imaginations, and community practices. However, because of the nature of twenty-first-century communication, these images are rarely seen in their aggregate form. Reported in disconnected frag-ments, a beautiful future is difficult to visualize or imagine. The mosaic pieces are scattered all over. What if students and the rest of us begin to perceive them as connected?

I find comfort in the words of Wendy Webber, a young writer for *State of Formation,* an online forum for emerging religious and ethical leaders. In 2018, she wrote, "If I am to have any chance to make a difference, my attitude has to be, 'we can't solve this today, but what can we do today so that it can be solved tomorrow?'"[176] Surely, our sense of loss can be trans-formed into creativity, good work, and future life, but the loss will never fully disappear, not for this generation or the next. How do we respond to the possibility that we may not see success in our lifetime, that life may well get worse before it gets better? Here is one answer. We don't quit; we notice deeply, we love, and we take action — no matter how small.

I often think of a young American who said to me, "I'm twenty years old, and all my life I've been told it's too late." Is it too late?

That may seem true, given the political dismissal of climate change, biodiversity collapse, inequity, and challenges to democracy. Right now it certainly seems so in my country. But what if democracy were revived in part through our relationship to the natural world? To develop a sense of the larger community, a child or an adult must step outside, must get to know the neighbors, both human and other-than-human.

What if nature were *widely* considered a human right — not only for the children of parents who love nature or who live in specific zip codes but for all children?

What if children were to learn of the joy of nature before they learn of its loss?

What if families, schools, businesses and government were to adopt a new equation: The more high-tech our lives become, the more nature we need.

What if for every education dollar we spend on the virtual, we spend at least another dollar on the real?

What if every family had the choice to send their child to a nature-based or forest school?

What if every playground were greened?

What if we naturalize our cities, homes, schools, and workplaces?

What if connecting people to nature became a growth industry?

What if leaders in every profession, every faith, every political party spent less time telling us what to do and more time asking us what we can create together?

What if, at the very moment when nature seems doomed, it saves us?

What if we dare to be idealistic once again?

The science-fiction vision of terraforming Mars — now seen as scientifically possible — imagines seeding a distant, lifeless planet. Now consider an alternative future: the terraforming of *Earth*. Imagine a gentle rain of green covering our fields and cities, seeds sprouting everywhere, wild animals returning, but this time for good.

What if? And why not?

I'M PRIVILEGED TO LIVE in these mountains and lucky to be reminded daily not only of the beauty and wonder of nature but also of its destruction by creatures just like me.

Transcendentalism was a nineteenth-century movement of writers and philosophers in New England, including Ralph Waldo Emerson and Henry David Thoreau, who sought to escape the industrial age by returning to nature.

Did Thoreau escape industrialism at Walden Pond? For a while, probably. Earth movers and other heavy machinery were not yet chipping away at those woods.

Today, every forest and every pond, including Walden Pond, is threatened by climate change and biodiversity collapse, and in direct consequence is the loneliness of our species. The loss is not an abstraction. When I moved here, I convinced myself that escape from the city was a form of transcendence. As it has turned out, I feel a greater sense of loss here than I did in the city. In the city, the loss of nature is muted by freeway noise and distraction. Here, we cannot escape it.

Every year, we see more old oak skeletons leaning, crouching, and then falling to their knees. Oak bones are scattered throughout the forest like tusks in an elephant graveyard. The life of the acorn woodpecker is linked to that of the oaks. The birds' habitat shrinks each year. As the oak groves vanish, so will the woodpecker dynasties, regardless of how much wealth they have stored in their granary banks, gutters, or vent pipes. Can we be a transcendentalist in a world with no escape from such threats? Experiencing nature brings us closer to beauty, wonder, and creation, but that closeness makes the ultimate threat even harder — because we become intimately aware of what we are losing.

Perhaps we need a new transcendentalism, a set of beliefs for this age — not an acceptance of destruction but a determination on both the global and the local scale not only to conserve what remains but also to create new natural environments.

When we consider our path to the future, we usually think of the way technology shapes what lies ahead, or we consider war, charismatic leaders, education, religion, or the comet that might come out of nowhere shattering the future. We think primarily about our own species, considering ourselves at the pinnacle of all life. We mainly wonder what will happen to us.

What would a future be like in which the welfare of all species was considered — a future in which we would recognize that we are part of nature, that all of nature is within us, and that we are part of a larger family?

In that future, we would shift from taking life to giving life. We would bring back the eastern ash, the whitebark pine, and the disappearing black oaks; we would revive the plummeting insect populations and the North American songbirds, whose numbers have fallen by a third over the past fifty years. In the meantime, our most effective antidote to ecological anxiety is noticing, cherishing, and protecting the wildness that survives within our animate circles.

Recently, I was walking on Boulder Creek Road, which winds down one of the southern flanks of Julian. Behind me, Volcan Mountain had exploded in color.

The air felt like a cool drink on a hot day.

During that walk, I enjoyed a conversation with a young gopher snake on a mission. It was the same species and similar in size to the snake I had seen flattened on another road two days earlier. I stood very still between this living snake and the other side of the road.

It began to curve around the back of my heel. Inching forth, it slowly moved the first five inches or so of its body forward and then tapped the shoe with its nose. It pulled away, then moved back, then forward again, and tapped three more times. The snake moved rhythmically around my foot.

We were on a busy road. A truck was speeding toward us. I reached down, touched the gopher snake on its back, and helped it cross the pavement. When we reached the edge, the snake moved away from the pavement and into the ground cover. It vanished, as if entering another time.

A Man Who Planted Trees

W HEN I WAS THREE or four, I was playing alone in the quince orchard outside my grandmother's house in Independence, Missouri. I was running around. It was cold, and I was wearing one of those hats with earflaps. I came to a little ditch and jumped over it. Memories are tricky, I know this. But in the middle of the jump, I froze in the air, looking down at the leaves submerged in clear water.

At that instant, when the world stopped, I had an overwhelming feeling that there was something I needed to do in life. Then the world returned, and I completed the jump. That's it.

But for the rest of my life, that memory would periodically return, every detail clear. The memory haunted me. I always associated it with my work. At every stage of my life, when I would do something — write a newspaper series, pen another book, whatever — I would ask myself, "Was that it? Did I do it?" Whatever I had done was never it. By the time I was in my fifties, I felt that I had failed. I had never done whatever it was I was supposed to do.

Then, in 2005, when *Last Child in the Woods* was published, I was on a morning radio show discussing the separation of children and the rest of us from nature, a deficit that many of us were beginning to recognize. And the calls started coming in. In that moment, as clear in its detail as the first memory, I knew I had done at least some of what I was supposed to do. But of course, the work had just begun.

I attribute much of what has been good in my life to nature's voice. Whether imagined or real, the voice is there; it rises from that sense of awe that the three- or four-year-old boy felt as he hovered over the ditch filled with wet leaves. Surely, the boy is still there, midair, listening.

Along with many others.

WHEN I WALK through these woods, I often wonder, "What are we looking for in nature and in places we choose to make our homes?" In an essay on why she became a writer and on the ability that an artist of any kind must have to absorb the shock of the world, Virginia Woolf wrote, "Behind the cotton wool is hidden a pattern...the whole world is a work of art...we are the words; we are the music; we are the thing itself."[177]

In their hypotheses, the artist, the scientist, and the priest ask the same fundamental questions. These are the questions I ask most when I walk in these hills: Who or what creates the patterns behind the wool? Are the patterns simply context or projection? What is it we're looking for? What is it that we yearn for, call for, search for in the natural world? Why there, and not in the steel girders of a high-rise condo? Surely whatever exists in the molecules of tempered steel or brick or stucco is — of course! — also nature. But most of us do not search there.

We search for *it* in the gaze of lovers and friends and children. Bigfoot hunters hope to look it in the eye. We search in forests and rivers and mountains and the little clump of trees at the end of the alley, in the dark entry points into whatever lies within and behind those trees. We search in the sky, too, in heaving cumuli rising from the desert to be collected and held by mountain peaks until they can be held no more.

Here is the mystery that has no name or thousands of names. *It* exists nowhere and everywhere, signaling through underground mycelia to young maples and ancient oaks, swirling through brown water to the reeds that hold glowing red clumps of dragonfly eggs and up the stalks to where blackbirds sing. It ferments in the soil.

It's in the seed of a dandelion, flowing on currents of air to places unknown. It's the whisper of leaves falling, it's in the tight little holes drilled by woodpeckers to store their faith in the future. Listen to the wind blow against and through the empty holes, the fluted sound of nature singing to itself. Does this song exist?

Of course it does, and of course the song is the music within us.

Here is my hypothesis: *It* exists because we want it to exist. Or because it wants us to exist. Within us or without us, it does not need our permission, approval, or proof to be. Without it, there is no beauty, no love. It arrives on lifting wings, the light behind the light.

CODY PETTERSON LOCKED his truck and shouldered his pack. We climbed a trail up a lower slope of Volcan Mountain through a patchwork of invasive grasses and ceanothus, crumpled skeletons of once-great oaks, the black spires of burned ponderosa, and infant pines. A red-tailed hawk balanced in an updraft. It was a cold day on the mountain. Cody moved quickly, ball cap pushed back on his head, long legs stretching from rock to boulder to trail. He is clearly bioenchanted.

With his family, as a boy, he bonded with these mountains, camping, hiking, and wandering, particularly on Volcan. Today, he is an educator and anthropologist. A few months after this trek, he was elected to the San Diego Unified School District board and is now its president. He lives on the coast, but his heart is rooted in these slopes.

In the drought years following the series of devastating fires, people who loved these mountains avoided them. They described the pain they felt, driving past Volcan, Cuyamaca, and the western slopes of Laguna, which retain barren scars, charred remains of trees and shrubs. Some former forests did gradually turn green, sometimes with new pines but usually with ceanothus bushes and invasive grasses that set up fuel for the fires next time. And there were more firestorms — not only here but also in Northern California, Colorado, New Mexico, Australia, and Canada. Once, these fires had been called forest fires or brush fires. As firefighter Jim Baker had told me, wildfires now often become superfires that produce their own weather, including towering firenadoes.

Cody could not bear to see Volcan's forests destroyed. For five years, usually a few days each week, he would drive for over an hour to Volcan with pine seedlings in his truck, dig holes, and gently plant the small trees. At first, he planted Bigcone Douglas-fir trees; later, he switched to Coulter pines.

He planted hundreds of trees on his own, with no fanfare.

On Earth Day 2019, after spending the day planting and watering seedlings, as he had for years, he wrote on Facebook, "It took months to figure out what was what, heading up to the mountain once a week, taking pictures, coming home and trying to identify all the species, reading late into the night about botany, and forestry."

He collected thousands of pinecones from the mountainside and learned how to extract seeds from them and how to "stratify, germinate, and pot the seeds" in his backyard. Putting his lawyerly skills to use, he assembled a working group with the US Forest Service, the US Fish and Wildlife Service, the California Department of Forestry and Fire Protection, and the US Department of Agriculture's Natural Resources Conservation Service. "We collected and sent 30 bushels of fresh cones up to the USFS nursery in Placerville," he wrote, "and I eventually got a thousand seedlings from those seeds," all from genetic stock specific to Volcan.

Today, Cody isn't the only one planting seedlings on Volcan Mountain.

Other organizations have the same aim. Bulldozers from some of the same agencies that Cody enlisted in his effort are currently clearing dead trees to reduce future fire hazard. No living trees are removed. But here and there, massive bald spots can now be seen on some of the slopes. The bulldozers and forestry teams are mainly removing massive tangles of monotypic ceanothus. The invasive plant had taken over the slopes and ravines after the two overlapping wildfires that raged through the backcountry two decades ago. The fires not only killed the trees but also destroyed the forest's ability to regenerate, because the intense heat killed the seeds and the next generation of saplings.

Eric Jones, president of the Volcan Mountain Foundation, explains, "Our forest restoration project removes the barriers to natural regeneration and reintroduces saplings — more than six thousand this next December with thousands more in coming years. We're bringing back the forest and habitat that was there before and returning a healthy resiliency. Remarkable stuff."

When Cody began replanting on his own, the process was complicated and discouraging. "I planted every which way I could," he told me. "The first year, I planted in the open. The seedlings baked." Afterward, he planted in the shade. He learned he had to water the young trees every two or three weeks, not an easy task across three hundred acres of steeply sloped terrain.

"The pocket gophers ate them from below," he said. "Rabbits severed them at the base. I caged the seedlings aboveground." Rodents ate the needles on all the branches that protruded from the cage, the hardware-cloth cages heated up in the sun, and the metal killed all the branches and needles that were in contact with it.

During all those months and years, the relentless heat and drought killed any seedlings he couldn't get to. Winter rains helped, but during the drought, there was no significant snowmelt.

And October rain doesn't help a seedling survive when there hasn't been a drop of rain in the previous eight months. Despite thousands of hours of work and care, Cody lost 650 of the 700 seedlings he'd raised and planted with his own hands over the previous five years.

And now, on our hike, he looked at the seedlings still held in his pack.

Cody continued: "One day, after a tough planting, I walked to one of my favorite spots, a ring of granite boulders sheltered by a giant, gnarled canyon live oak." There, lying shattered and rotting in the middle of the ring, was half of a sixty-foot-tall tree. The other half was still standing, but it was covered in the telltale tiny D-shaped exit holes of the goldspotted oak borer (GSOB), a beetle whose larvae feed beneath the bark near the interface of the phloem and xylem tissues, where nutrients and water are conducted.

The larvae also damage a layer responsible for the radiating growth of the tree. After a few generations of larvae do their thing, the tree dies. Since 2008, the beetle has killed more than eighty thousand native oaks in San Diego County, starting with the largest, most magnificent giants.[178]

That day of realization, Cody walked among the ancient oaks in his favorite spot on the mountain. He saw that every trunk was spotted with GSOB larvae holes. "I stood there stunned. The whole millennia-old forest was dying, as far as the eye could see. I wandered back to my truck, numb. I sat down in the driver's seat, staring out the window. At the oaks, dying."

He thought about the hundreds of pines he had planted on this mountain and the hundreds that had died. He remembered the seedlings he planted to replace them. "I've lost too many to count, but I can still recall the moment I first saw each one that had dried out, been stripped bare, pulled under by gophers, gnawed by rabbits, or trampled by cattle from the neighboring reservation. I've thought about it all a thousand times. I've lain in bed so many nights trying to wrestle with it. That afternoon, something in my mind buckled under the weight of it. I thought, 'How do I tell my kids?' and for the first time in fifteen years, I started to cry."

The obvious doubled him over.

"And then I had the saddest thought I've ever had: 'I wish I didn't know.'"

In the quiet that settled around us, the fragile seedling stood as a silent testament to a faith neither of us could fully name. There was no sound except the wind and the distant hum of insects. Cody pressed the soil around the fragile stem, his hands moving with a gentle ritual, as if shielding something ancient and sacred.

"I guess this is the only answer I have," he murmured. "You keep planting, loving, tending, because that's what the Totality asks of you. Not because you expect to save it all, but because you know, in your bones, that each small act is a thread in something vast."

For a long moment, we listened to the mountain's breath — the hush of needles, the faint, living pulse beneath our knees. There was no guarantee these trees would survive the following summer or outlast the fires that loomed ever closer each year. Still, Cody rose, brushed the dark soil from his palms, and smiled with a hard-earned grace that held both weariness and hope.

That day with Cody, I remembered a question that had been nagging me for weeks: How do we love nature when so much of it is being destroyed around us?

"Don't mourn," he said. "Plant."

*　*　*

WE LOVE NATURE while we can in the ways that we can.

We care for the climate refugees, the animals and plants immigrating into our area and those relocating to northern climates. We love nature by respecting and maximizing its resilience. We promote more-natural environments, plant native species, support the rewilding of cities, and practice regenerative agriculture. We create animal overpasses and wildlife corridors to aid animal migrations through urban areas or to connect fragmented wild lands.

Or we love nature by trying to preserve it exactly the way we think it is. Which seldom works, because humans and the rest of nature must learn to be resilient, to change, even now.

And what if it's too late?

When a friend, a mother, a father, a daughter, a son, or a grandmother — or nature itself — begins to fade from this life, we do not walk away. In the final months, weeks, days, and hours, we stay by the side of that loved one. We hold their hand, and they have ours. We listen to their stories, their voice, and, finally, their breath. During this time of metamorphosis, we come to know them more deeply than we ever did when we believed that life was unending. Secrets reveal themselves. History becomes clear in that light.

Becoming gripped by guilt over past sins is no help to anyone. Love is the hope. If we have failed at love in the past, we now rise to the challenge. Even if only to feel less alone, we can turn to those other lives and to each other. We do need a new spirit.

I asked Cody what still motivates him. He had difficulty answering that question.

"There's this beautiful passage in the New Testament," he said. "I'm forgetting what it is right now, but the apostle Paul says the whole of the law is one command. To love thy neighbors as thyself. You ask yourself, how is it that the whole of the law could be one commandment, to love your neighbor?"

"These trees, all of these, are your neighbors?" I said.

"Exactly. I am identified with this system I love," he said. "I mean, you could say, like, 'Oh, it's okay that the forest burned down. Now it's ceanothus's turn. Ceanothus is great, too, right?' And there are folks at the state level who say we shouldn't clear it. Let ceanothus be. Ten thousand years from now, ceanothus will be gone, and the forest will climax to something else.

"But that's not for me. I'm attached to the system of life that I've come to know."

The Dutch artist Jan van Boeckel also struggles with these questions. He suggests the practice of mindful schizophrenia, which refers to the idea "that we cannot escape living with a 'split mind,' loving nature while acknowledging the impact of our inevitable ecological footprint." Rather than denying or ignoring this condition, he advises, we should try to embrace it gracefully and mindfully and do what we can to nurture the life around us.

Earlier that day, Cody had said he took no pleasure in planting. But now he looked up at a slope of new young trees that he had planted.

"You're taking pleasure right now, Cody," I said. "It's in your face."

"Yeah, I guess that's right," he said. "I certainly take a certain amount of pride in it. I don't know where it comes from. I have a sense of—it's not pleasure, but it's like a sense of well-being and fulfillment and being, almost, like it's an out-of-body…it's like you feel you're being moved by something—almost like a puppet, but the thing you become is like…the Totality.

"I was talking to someone yesterday who's a young activist. I said, 'Look, you have to behave as though your body is a disguise or a costume that the Totality is wearing to walk amongst us.'"

Totality is what he calls the life force, God. He has no other name. It.

Cody reached into his pack and gently pulled out another seedling. He knelt on the earth and began to dig a hole with his hand. I asked him again why he continued to plant trees. He bent his head down. It took him a long time to find his voice, and it was broken. He said something about fealty to the Totality, which he feels every day in those trees, old and new.

His explanation deals directly with the sense of despair that so many people now feel and why it's important to keep caring for nature, just as we would a grandmother, no matter what — though we know the shortness of life. Between now and when the smoke rolls over the mountain again and the flames rise, life on this mountain will be made a little better for the living. Even if only for a while. Cody Petterson suffered his crises five years ago. But that moment of despair was not the end of his story.

He's been planting trees ever since.

ACKNOWLEDGMENTS

NO BOOK IS conceived or delivered alone.

I am forever indebted to my wife, Kathy, who helped make this book possible at every stage and at every level, and to my sons, Matthew and Jason, who by words and example offered keen insights and encouragement. And to Algonquin Books, my publisher for twenty years—and especially my patient and talented editors, Madeline Jones and Amy Gash—along with Brunson Hoole, Arik Hardin, June Park, Sally Kim, Elizabeth Johnson, and Patricia E. Boyd. My literary agent James Levine helped keep this book on track, as did Jackie Green, who for more than three decades has handled my public sermonettes.

Dean Stahl, Janet Keating, and other friends contributed more than they know. So did John Johns, Bob Schneider, Chris Khoury, Linda Corey, Larry Hinman, Ron and Janice Swaisgood, Cheryl Charles, Mike Louv, and Amy Pertschuck. Also, my gratitude to Jan van Boeckel, Patti Shields, Kathleen Lockyer, Pat Hasbach, and many others—including fellow journalist Cathy Scott, who shared commiseration and encouragement at the the Purple Owl Cafe. My heartfelt appreciation also goes to Sam Schultz and Robyn Bjornsson.

I am also grateful to everyone at the Children & Nature Network (C&NN), the nonprofit that emerged from *Last Child in the Woods*. C&NN continues to help people, organizations, and governments connect children, families, and communities to the natural world. Also, to the Volcan Mountain Foundation, which for decades has been dedicated to protecting the land and magic of this region. And of course, *Noticing* could not have been written without the good people in these mountains who shared their stories, humor, wisdom — and their love for this bioenchanted place.

SUGGESTED READING

Abram, David. *The Spell of the Sensuous: Perception and Language in a More-Than-Human World*. Vintage, 1997.

Ackerman, Diane. *A Natural History of the Senses*. Vintage, 1991.

Albrecht, Glenn. *Earth Emotions: New Words for a New World*. Cornell University Press, 2019.

Anderson, Doug, Lorraine Chiarotto, and Julie Comay. *Natural Curiosity: A Resource for Educators — Considering Indigenous Perspectives in Children's Environmental Inquiry*, 2nd ed. Ontario Institute for Studies in Education, 2017.

Anderson, M. Kat. *Tending the Wild: Native American Knowledge and the Management of California's Natural Resources*. University of California Press, 2013.

Apelian, Nicole. *The Forager's Guide to Wild Foods*. Global Brother, 2021.

Arbabi, Sean. *The Complete Guide to Nature Photography*. Amphoto Books, 2011.

Banyard, Ian, and Miura Miyuki. *Natural Mindfulness: Your Personal Guide to the Healing Power of Nature Connection*. Vision Maker Press, 2018.

Bekoff, Marc. *Rewilding Our Hearts: Building Pathways of Compassion and Coexistence*. New World Library, 2014.

Berman, Marc. *Nature and the Mind: The Science of How Nature Improves Cognitive, Physical, and Social Well-Being*. S&S/Simon Element, 2025.

Berger, John. *About Looking.* Pantheon Books, 1980.

Berry, Thomas. *The Dream of the Earth.* Counterpoint Press, 2015.

Burghardt, Gordon M. *Genesis of Animal Play: Testing the Limits.* MIT Press, 2005.

Carrico, Richard. *Strangers in a Stolen Land.* Sunbelt, 2008.

Carson, Rachel. *The Sense of Wonder.* Harper & Row, 1956.

Charney, Noah, and Charley Eiseman. *Tracks & Sign of Insects and Other Invertebrates.* Stackpole Books, 2010.

Chawla, Louise. *In the First Country of Places: Nature, Poetry, and Childhood Memory.* SUNY Press, 1994.

Childs, Craig. *The Animal Dialogues: Uncommon Encounters in the Wild.* Back Bay Books, 2007.

Daston, Lorraine, and Katharine Park. *Wonders and the Order of Nature, 1150–1750.* Zone Books, 2001.

Farmer, Jared. *Elderflora: A Modern History of Ancient Trees.* Basic Books, 2022.

Fazekas, Andrew. *National Geographic Backyard Guide to the Night Sky,* 2nd ed. National Geographic Books, 2019.

Godfrey-Smith, Peter. *Other Minds: The Octopus, the Sea, and the Deep Origins of Consciousness.* Farrar, Straus & Giroux, 2016.

Gonzales, Laurence. *Everyday Survival: Why Smart People Do Stupid Things.* W. W. Norton, 2009.

Goodall, Jane, and Douglas Abrams. *The Book of Hope: A Survival Guide for Trying Times.* Macmillan, 2021.

Gooley, Tristan. *The Lost Art of Reading Nature's Signs.* The Experiment, 2015.

Greene, Harry Walter, and Michael Fogden. *Snakes: The Evolution of Mystery in Nature.* University of California Press, 1997.

Harding, Stephan. *Animate Earth. Science, Intuition, and Gaia.* Green Books, 2006.

Hasbach, Patricia H. *Grounded: A Guided Journal to Help You Reconnect with the Power of Nature — and Yourself.* Adams Media, 2022.

Hasbach, Patricia H. *Prescribing Nature: A Clinician's Guide to Ecotherapy.* W. W. Norton, 2025.

Henshaw, John M. *How Many Senses Do We Have?* Johns Hopkins University Press, 2012.

Higgs, John. *William Blake vs the World.* Hachette UK, 2021.

Hutto, Joe. *Illumination in the Flatwoods.* Lyons Press, 2006.

Hutto, Joe. *Touching the Wild: Living with the Mule Deer of Deadman Gulch.* Skyhorse, 2014.

Jung, Carl G. *Memories, Dreams, Reflections.* Pantheon Books, 1963.

Kahn, Peter H. Jr., and Patricia H. Hasbach. *Ecopsychology Science, Totems, and the Technological Species.* MIT Press, 2012.

Kaplan, Rachel, and Stephen Kaplan. *The Experience of Nature: A Psychological Perspective.* Cambridge University Press, 1989.

Kaur, Valarie. *See No Stranger.* One World, 2021.

Kershenbaum, Arik. *Why Animals Talk: The New Science of Animal Communication.* Penguin Press, 2024.

Kimmerer, Robin Wall. *Braiding Sweetgrass: Indigenous Wisdom, Scientific Knowledge, and the Teaching of Plants.* Milkweed Editions, 2013.

Krause, Bernie. *Wild Soundscapes: Discovering the Voice of the Natural World.* Yale University Press, 2016.

Lawrence, Natalie. *Planta Sapiens: The New Science of Plant Intelligence.* W. W. Norton, 2023.

Laws, John Muir. *The Laws Guide to Nature Drawing and Journaling.* Heyday Books, 2016.

Leopold, Aldo. *Sand County Almanac and Sketches Here and There.* Oxford University Press, 1972.

Li, Dr. Quing. *Forest Bathing: How Trees Can Help You Find Health and Happiness.* Penguin Life, 2018.

Lockyer, Kathleen. *Wild Inside: How Nature Protects Your Child's Mental Health and Restores Yours.* Otter Play Publishing, 2025.

Long, William J. *How Animals Talk: And Other Pleasant Studies of Birds and Beasts.* Bear, 2005.

Louv, Richard. *Greening Democracy: The Transformative Power of Nature on Children and Society.* The MIT Press Reader, January 12, 2024.

Louv, Richard. "Outdoors for All: A Nascent Global Movement Proclaims That Access to Nature Is a Human Right." *Sierra Magazine,* May 7, 2019.

MacFarlane, Robert. *Is a River Alive?* W. W. Norton, 2025.

Marcus, Clare Cooper, and Naomi A. Sachs, *Therapeutic Landscapes: An Evidence-Based Approach to Designing Restorative Outdoor Spaces.* Wiley, 2013.

Marshall, Albert D., and Louise Zimanyi. *Walking Together.* Annick Press, 2023.

McGinnis, Michael Vincent. *Science and Sensibility: Negotiating an Ecology of Place.* University of California Press, 2016.

McKibben, Bill. *Earth: Making a Life on a Tough New Planet.* Times Books, 2010.

McNight, Carl B., et al. *Peterson Field Guide to Mushrooms of North America.* New York: Mariner Books, 2021.

Monbiot, George. *Feral: Rebuilding the Land, the Sea, and Human Life.* University of Chicago Press, 2014.

Moore, Robin C., and Herbert H. Wong. *Natural Learning: The Life History of an Environmental Schoolyard.* MIG Communications, 1997.

Nygren, Steve. *Start in Your Own Backyard: Transforming Where We Live with Radical Common Sense.* Matt Holt Books, 2025.

Orr, David W. *Earth in Mind: On Education, Environment, and the Human Prospect.* Island Press, 1994.

Plotkin, Bill. *Soulcraft: Crossing into the Mysteries of Nature and Psyche.* New World Library, 2003.

Pretor-Pinney, Gavin. *The Cloudspotter's Guide: The Science, History, and Culture of Clouds.* Tarcher, 2007.

Pyle, Robert Michael. *The Thunder Tree: Lessons from an Urban Wildland.* Houghton Mifflin, 1993.

Rezendes, Paul. *Tracking and the Art of Seeing: How to Read Animal Tracks and Sign.* HarperCollins, 1999.

Sibley, David Allen. *What It's Like to Be a Bird: From Flying to Nesting, Eating to Singing — What Birds Are Doing, and Why.* Knopf, 2020.

Safina, Carl. *Beyond Words: What Animals Think and Feel.* Souvenir Press, 2018.

Schlanger, Zoe. *The Light Eaters: How the Unseen World of Plant Intelligence Offers a New Understanding of Life on Earth.* Harper, 2024.

Simard, Suzanne. *Finding the Mother Tree: Discovering the Wisdom of the Forest.* Vintage, 2021.

Strycher, Noah. *National Geographic Birding Basics.* National Geographic Books, 2022.

Tan, Amy. *The Backyard Bird Chronicles.* Knopf, 2024.

Toft, Roy. *Wild Ramona.* Roy Toft, 2023.

Toler, David L. Jr. *Blood of the Band: An Ipai Family Story.* Sunbelt, 2015.

Toone, Bill. *On the Wings of a Condo.* Ecolife Conservation, 2023.

Van Boeckel, Jan. *At the Heart of Art and Earth: An Exploration of Practices in Arts-Based Environmental Education.* Aalto University, 2013. (Also available as a free downloadable PDF at https://shop.aalto .fi/media/attachments/0563b/vanboeckel.pdf.)

Wilken-Robertson, Michael, Deborah Small, et al. *Kumeyaay Ethnobotany: Shared Heritage of the Californias.* Sunbelt, 2017.

Wilson, Edward O. *Biophilia.* Harvard University Press, 1984.

Wohlleben, Peter. *The Hidden Life of Trees: What They Feel, How They Communicate — Discoveries from a Secret World*. Greystone Books, 2016.

Wohlleben, Peter. *The Heartbeat of Trees — Embracing Our Ancient Bond with Forests and Nature*. Greystone Books, 2021.

Wulf, Andrea. *The Invention of Nature: Alexander von Humboldt's New World*. Vintage Books, 2015.

Young, Jon, and Dan Gardoqui. *What the Robin Knows: How Birds Reveal the Secrets of the Natural World*. Mariner Books, 2012.

Young, Jon, Ellen Haas, and Evan McGown. *Coyote's Guide to Connecting with Nature*. OWLink Media, 2010.

Additional Resources

Reader's guides for *Noticing, Our Wild Calling, Vitamin N, The Nature Principle, Last Child in the Woods*: www.richardlouv.com.

The Children & Nature Network Research Library: Abstracts of over 1,500 studies on the benefits of nature connection to the physical, mental, and cognitive health of children and adults: www .childrenandnature.org/learn/research/.

Substack: https://substack.com/@richardlouv.

For more about Richard Louv and his books and newsletter, visit www.richardlouv.com.

NOTES

Introduction

1 **Having never seen a garter snake:** Joe Hutto, *Illumination in the Flatwoods* (Lyons Press, 2006).

2 **Today, over fifteen hundred:** Children & Nature Network Research Library, www.childrenandnature.org/learn/research/.

3 **In 2025:** Helen Coffey, "Touch Grass: How Gen Z Stopped Going Outside and Why It Matters," *Independent*, September 19, 2025.

Chapter 2

4 **"If you would make acquaintance":** Henry David Thoreau, *The Journal 1837–1861*, ed. Damion Searls (New York Review Books Classics, 2009).

5 **The researchers' conclusion:** Becky Schauer, Kishori Koch, Laura Lemieux, and Kendra Willey, "How Immersion in Nature Impacts the Human Spirit: A Phenomenological Study" (master's thesis, St. Catherine University, 2016), https://core.ac.uk/download/pdf/200292511.pdf.

6 **D. H. Lawrence goes further:** D. H. Lawrence, *Etruscan Places* (Leiserson Press, 2007).

7 **Once, the valley held:** Robert Sheehan, "Water Rights, Economics, and Displacement: The Kumeyaay Story," Digital History Project, California State University San Marcos, https://wp.csusm.edu/rsheehan/capitan -grande-act/.

8 **Some of them may have been:** Carl Zimmer, "Humans Lived in North America 130,000 Years Ago, Study Claims," *New York Times*, April 26, 2017, www.nytimes.com/2017/04/26/science/prehistoric-humans-north -america-california-nature-study.html.

9　**The website of the Viejas Band:** Viejas, "Geology and Minerals," https://viejasbandofkumeyaay.org/viejas-community/kumeyaay-history/geology-and-minerals/.

10　**In *Fremontia*:** Thomas A. Oberbauer, "Sierra de San Pedro Martir," *Fremontia: Journal of the California Native Plant Society* 24, no. 4 (1996): 21–24.

11　**Plants and animals:** Volcan Mountain Foundation, www.volcanmt.org/plants-animals.

12　**In stillness:** Jenny Odell, *How to Do Nothing: Resisting the Attention Economy* (Melville House, 2019).

13　**"sum up in a simple word":** Jason M. Brown, *Dwelling in the Wilderness* (Trinity University Press, 2024).

14　**"many pathways we might take":** Sarah Elisa Kelly, "Practising Place-Fullness," *Resilience*, February 17, 2021, www.resilience.org/stories/2021-02-17/practising-place-fullness/.

15　**For "place bonding":** Patricia Hasbach, *Prescribing Nature: A Clinician's Guide to Ecotherapy* (W. W. Norton & Company, 2025).

16　**people would not be so "divided":** Carl G. Jung, *Memories, Dreams, Reflections* (Pantheon Books, 1963).

17　**Nearly 9 million animal species:** See, for example, John J. Wiens, "How Many Species Are There on Earth? Progress and Problems," *PLOS Biology* 21, no. 11 (2023), https://doi.org/10.1371/journal.pbio.3002388.

18　**In 2024, the PBS series *Nature*:** Amanda Schmidt, "San Diego: America's Wildest City," *Nature*, November 6, 2024, www.pbs.org/wnet/nature/about-san-diego-americas-wildest-city/31567/.

19　**It's home to "200 imperiled plants and animals":** The Nature Conservancy, "San Diego Is the Most Biologically Rich County in the Continental U.S.," www.nature.org/en-us/get-involved/how-to-help/places-we-protect/san-diego-county/.

20　**more Indigenous tribes and reservations:** University of San Diego, "Indian Reservations in San Diego County," www.sandiego.edu/native-american/reservations.php.

Chapter 3

21　**In his book:** Gary Snyder, *The Practice of the Wild* (Counterpoint, 2010).

22   **"presence of something":** Dacher Keltner, *Awe: The New Science of Everyday Wonder and How It Can Transform Your Life* (Penguin, 2024).

23   **But look in the mirror:** Diane Ackerman, *A Natural History of the Senses* (Phoenix, 1990).

Chapter 4

24   **Just add water:** "Cryptobiotic Soil Crusts — Glen Canyon National Recreation Area (US National Park Service)," National Park Service, www.nps.gov/glca/learn/nature/soils.htm.

25   **an alternative term:** Kathryn M. Parsley, "Plant Awareness Disparity: A Case for Renaming Plant Blindness," *Plants, People, Planet* 2, no. 6 (2020), https://doi.org/10.1002/ppp3.10153.

26   **"to recognize the role":** Ainara Achurra, "Plant Blindness: A Focus on Its Biological Basis," *Frontiers in Education* 7 (October 25, 2022), https://doi.org/10.3389/feduc.2022.963448.

27   **Research reveals:** David R. Hershey, *Plant Content in the National Science Education Standards,* Action Bioscience, https://files.eric.ed.gov/fulltext/ED501357.pdf

28   **"change their visual signal":** Achurra, "Plant Blindness."

29   **Most people who do not see plants:** Jonathon Keats, "A Clock in the Forest," *NOĒMA,* February 8, 2024, www.noemamag.com/a-clock-in-the-forest/.

Chapter 5

30   **The study found these potential benefits:** Ye Wen et al., "Medical Empirical Research on Forest Bathing (Shinrin-Yoku): A Systematic Review," Environmental Health and Preventive Medicine 24, no. 70 (2019), https://doi.org/10.1186/s12199-019-0822-8.

31   **sandwich, reportedly marked with the image of the Virgin Mary:** "'Virgin Mary' Toast Fetches $28,000," BBC News, November 23, 2004, http://news.bbc.co.uk/2/hi/4034787.stm.

Chapter 6

32   **"The Middle English word":** John Higgs, *William Blake vs the World* (Hachette UK, 2021).

33   **a children's book about how to practice etuaptmumk:** Albert D. Marshall and Louise Zimanyi, *Walking Together* (Annick Press, 2023).

34   **"With my two eyes I see everything":** Erin Pottie, "Eskasoni Elder Bringing 'Two-Eyed Seeing' into Classrooms," *CBC News,* March 18, 2022, www.cbc.ca/news/canada/nova-scotia/eskasoni-elder-bringing-two-eyed -seeing-into-classroom-1.6388379.

35   **"The tree which moves":** William Blake to Rev. John Trusler, 1799. *The Complete Poetry & Prose of William Blake* (Anchor, 1982).

36   **Higgs writes:** Higgs, *William Blake vs the World.*

37   **"It was dark and dusty":** Laurence Gonzales, *Everyday Survival: Why Smart People Do Stupid Things* (W. W. Norton, 2009).

38   **"Some synesthetes hear":** Siri Carpenter, "Everyday Fantasia: The World of Synesthesia," *Monitor on Psychology,* March 2001, www.apa.org /monitor/mar01/synesthesia.html.

39   **Henshaw cautions:** John M. Henshaw, "How Many Senses Do We Have?" (Johns Hopkins University Press, 2012), www.press.jhu.edu/newsroom /how-many-senses-do-we-have.

40   **Quoted in *Monitor on Psychology*:** "Green Is Good for You," *Monitor on Psychology,* April 2001, www.apa.org/monitor/apr01 /greengood.

41   **While recording her teaching process:** Carol Black, "On the Wildness of Children: The Revolution Will Not Take Place in the Classroom," *Films for Action,* May 17, 2017, www.filmsforaction.org/articles/on-the-wildness -of-children/.

42   **"Looking beyond the central scene is key":** Frans de Waal, *Mama's Last Hug: Animal Emotions and What They Teach Us About Ourselves* (W. W. Norton & Company, 2019), 2.

43   **"I have huge problems":** Oliver Sacks, "Face-Blind," *The New Yorker,* August 23, 2010, www.newyorker.com/magazine/2010/08/30/face-blind.

Chapter 7

44   **According to the multidisciplinary journal:** "Mysterianism Lite," editorial, *Nature Neuroscience* 3, no. 3 (2000), https://doi .org/10.1038/72893.

45   *Thin place* **is an "odd term":** Eric Weiner, "Where Heaven and Earth
Come Closer," Cultured Traveler, *New York Times,* March 9, 2012, www
.nytimes.com/2012/03/11/travel/thin-places-where-we-are-jolted-out-of
-old-ways-of-seeing-the-world.html.

46   **Northern Irish writer Kerri ní Dochartaigh:** *Thin Places: A Natural
History of Healing and Home* (Milkweed Editions, 2023).

Chapter 8

47   **Wildlife conservationist Gautam Shah and novelist Ryan Boudinot:**
"Using the Metaverse to Connect and Protect the Natural World," *Stanford
Social Innovation Review,* July 5, 2022, https://ssir.org/articles/entry/using
_the_metaverse_to_connect_and_protect_the_natural_world.

48   **Indeed, recent research:** The human relationship with nature and
technological nature, The Children & Nature Network Resource Library,
https://research.childrenandnature.org/research/technological-nature
-may-be-better-than-no-nature-but-substituting-technological-nature-for
-real-nature-raises-concerns/.

49   **John Mack, an artist and photographer:** Life Calling, "A Species Between
Worlds," 2016, https://life-calling.org/initiatives/a-species-between
-worlds/.

50   **"And you know":** Jeanette Winterson, "Sex Bots, Religion and the Wild
World of A.I.," editorial, *New York Times,* November 1, 2021, www.nytimes
.com/2021/11/01/opinion/sway-kara-swisher-jeanette-winterson.html.

51   **Coincidentally:** Samantha Murphy Kelly, "I Tried Apple's New
Headset. Here's What It's Like to Use," CNN, June 6, 2023, www.cnn
.com/2023/06/06/tech/apple-vision-pro-hands-on/index.html.

52   **Such a tree:** Jared Farmer, *Elderflora: A Modern History of Ancient Trees*
(Basic Books, 2022).

53   **We are not the only time travelers:** Stanley Coren, "Can Dogs Smell
Time?" *Psychology Today,* November 22, 2019, www.psychologytoday.com
/us/blog/canine-corner/201911/can-dogs-smell-time.

54   **How differently?:** David Borkenhagen, "Octopus Time," *Aeon,* April
20, 2023, https://aeon.co/essays/can-the-liquid-motion-of-the-octopus
-radicalise-our-ideas-about-time.

55   **Marcia Bjornerud:** Princeton University Press, "Marcia Bjornerud on *Timefulness*," interview, Princeton University Press, July 10, 2018, https://press.princeton.edu/ideas/marcia-bjornerud-on-timefulness.

56   **"The memory of the world":** Sarah Hatfield, "Today You Are as Old as You Have Ever Been," Audubon Community Nature Center, December 28, 2016, https://auduboncnc.org/today-you-are-as-old-as-you-have-ever-been.

Chapter 9

57   **In the book:** Ian Banyard, *Natural Mindfulness: Your Personal Guide to The Healing Power of Nature Connection* (Vision Maker Press, 2018).

58   **Unable to hear or see:** Helen Keller, *The World I Live In* (The Century Co., 1908).

59   **Writing in *InnerVoices*:** Jude Colin, "The Forgotten Sense: How to Cultivate a Deeper Connection with Nature Through Touch," in *InnerVoices: Empowerment and Personal Development for Introverts* (blog), *Medium*, December 11, 2024, https://medium.com/happier-with-each-passing-day/the-forgotten-sense-how-to-cultivate-a-deeper-connection-with-nature-through-touch-dc22887e6045.

60   **Peter Wohlleben:** Peter Wohlleben, *The Heartbeat of Trees: Embracing Our Ancient Bond with Forests and Nature* (Greystone Books, 2021).

61   **A reporter for *Outside*:** Calvin Van Paris, "This App Wants You to Touch Grass — Literally," *Outside,* June 7, 2025, www.outsideonline.com/health/wellness/touch-grass-app/.

62   **Then there's *grounding*:** Julia Belluz, "The Truth About WebMD, a Hypochondriac's Nightmare and Big Pharma's Dream," *Vox,* April 5, 2016, www.vox.com/2016/4/5/11358268/webmd-accuracy-trustworthy.

63   **"natural electric charge of the earth":** Kristen Fischer, "Grounding: Techniques and Benefits," WebMD, May 3, 2024, www.webmd.com/balance/grounding-benefits.

64   **Lyall Watson:** Lyall Watson, *Supernature* (Coronet Books, 1999).

Chapter 10

65   **Her first feature-length film:** "Fox Maxy, Filmmaker & Visual artist," *Chanel* magazine, https://www.chanel.com/us/chanel-next-prize/2024/fox-maxy.

66   **In a 2025 article:** Fox Maxy, "Grief Doesn't End or Stop. But There's a Comfort in That," *Los Angeles Times,* March 17, 2025, www.latimes.com /lifestyle/image/story/2025-03-17/fox-maxy-on-grief.

67   **In his book:** Menno Schilthuizen, *The Urban Naturalist: How to Make the City Your Urban Playground* (MIT Press, 2025), xiv.

68   **"To develop a complete mind":** Katie White, "7 Scientific Pioneers Who Were Also Artistic Visionaries, from the Inventor of the Morse Code to the Founder of Neurobiology," *Artnet,* April 6, 2020, https://news.artnet.com /art-world/artists-who-were-scientists-1821906.

69   **"I could imagine the cells in there":** Richard Feynman, "Richard Feynman: Ode on a Flower," YouTube video posted by RichardFeynmanLove, October 12, 2010, www.youtube.com /watch?v=hIZhgLKSBaY.

70   **In her own ode to a flower:** Elizabeth Hutton Turner, *Georgia O'Keeffe: The Poetry of Things* (Yale University Press, 1999).

71   **My mother and Fox Maxy:** "The unexpected outcomes of artist-scientist collaborations," Nature Careers Podcast, *Nature,* Nov. 10, 2023, https:// www.nature.com/articles/d41586-023-03390-y.

72   **"*Everybody* was talking about it":** Richard Whittaker, "A Life of One's Own: A Conversation with James Hubbell," *Works & Conversations,* April 18, 2002, www.conversations.org/story.php?sid=57. (Emphasis in original.)

Chapter 11

73   **In 2022, he cofounded:** Wild Wonder Foundation, "Wild Wonder Foundation," 2019, www.wildwonder.org.

74   **That year, as she writes:** Amy Tan, "The Backyard Bird Chronicles," https://amytan.net/backyard-bird-chronicles.

75   **Later, she assembled her journals:** Amy Tan, *The Backyard Bird Chronicles* (Knopf, 2024).

76   **"keep engaged":** John Muir Laws, *The Laws Guide to Drawing Birds.* (Berkeley, California, Heyday, 2012).

77   **Paying attention:** *Nature Journaling with Expert John Muir Laws,* FOSS podcast, https://www.foss-science.com/nature-journaling/.

78   **In their book:** Lorraine Daston and Katharine Park, *Wonders and the Order of Nature, 1150–1750* (Zone Books, 2001).

Chapter 12

79 **"In the late Spring":** Portions of Susan McDonnell's email letter were published in Richard Louv, "The Natural Art of Seeing," Children & Nature Network, April 2020, www.childrenandnature.org/resources/the-natural-art-of-seeing/.

80 **In her book:** Clare Walker Leslie, *The Nature Connection: An Outdoor Workbook for Kids, Families, and Classrooms* (Storey Publishing, 2010), 30.

81 **The nineteenth-century Postimpressionist painter:** Nina Athanassoglou-Kallmyer, "'Like a Dog, Just Looking' — Cézanne, Innocence, and Early Phenomenological Thought in Nineteenth-Century France," *Société Cezanne,* June 27, 2019, www.societe-cezanne.fr/2019/06/27/like-a-dog-just-looking/.

Chapter 13

82 **"It's not the brain that sees":** Amanda Gefter, "What Plants Are Saying About Us," *Nautilus,* March 7, 2023, https://nautil.us/what-plants-are-saying-about-us-264593/.

83 **"Photography not only exposes":** Miles Richardson et al., "Actively Noticing Nature (Not Just Time in Nature) Helps Promote Nature Connectedness," *Ecopsychology* 14, no. 1 (2021), https://doi.org/10.1089/eco.2021.0023.

84 **Watkins's photographs:** Wilderness Society, "11 Incredible Wilderness Photographs from the 1800s," *Medium,* September 8, 2015, https://medium.com/@wilderness/11-incredible-wilderness-photographs-from-the-1800s-ffaa60d381c0.

85 **In his autobiography:** Ansel Adams, *Ansel Adams: An Autobiography* (Little, Brown, 1985).

86 **"Just by being out":** Melissa Groo, "Statement of Ethics," Melissa Groo Photography, 2023, www.melissagroo.com/clients/ethics/.

87 **"Starvation Mountain":** Roy Toft, *Wild Ramona* (Roy Toft, 2023).

88 **Here, the Viejas Band:** Ernie Cowan, "Secretive, Elusive Bobcats Are Similar to Spirit Animals," *San Diego Union-Tribune,* June 10, 2023, www.sandiegouniontribune.com/2023/06/10/secretive-elusive-bobcats-are-similar-to-spirit-animals/.

89   **WildCat Ridge Sanctuary:** "Bobcats," https://wildcatridgesanctuary.org
/residents/bobcats/.

Chapter 14

90   **Nick Romero:** Nick Romero, "Nicholas Humphrey's Beautiful Theory of
Mind," *The New Yorker,* March 15, 2023, www.newyorker.com/news
/annals-of-inquiry/nicholas-humphreys-beautiful-theory-of-mind.

91   **Smell is the sense:** Sara Tsompanidi, "The Extraordinary Relation Between
Our Senses and Our Memories," *Medium,* October 24, 2019, https://
medium.com/swlh/the-extraordinary-relation-between-our-senses-and
-our-memories-eb48be7ee136.

92   **"As humans, [we] have a discrete bandwidth":** Lee Burton, "What Is Bird
Language?" *Naturalist Studies,* May 24, 2025, www.naturaliststudies.com
/newsletters/naturalist-notes/posts/what-is-bird-language.

93   **According to the Cornell Lab of Ornithology:** Cornell Lab, "How to
Recognize Woodpeckers by Their Drumming Sounds," All About
Birds, April 3, 2025, www.allaboutbirds.org/news/how-to-recognize
-woodpeckers-by-their-drumming-sounds/.

94   **Over two hundred species:** Eric R. Schuppe et al., "Forebrain Nuclei
Linked to Woodpecker Territorial Drum Displays Mirror Those That
Enable Vocal Learning in Songbirds," *PLOS Biology* 20, no. 9 (2022),
https://doi.org/10.1371/journal.pbio.3001751.

95   **But woodpeckers coordinate:** Corrie Pikul, "Why Do Woodpeckers Peck?
New Discovery About Bird Brains Sheds Light on Intriguing Question,"
News from Brown (Brown University), September 20, 2022, www.brown
.edu/news/2022-09-20/woodpeckers-0.

96   **While the association:** American Birding Association, "ABA Recording
Rules & Interpretations" (version 2020), www.aba.org/aba-recording-rules
-and-interpretations.

97   **particularly after the pandemic:** Katinka Domen, "ABA Recording Rules
& Interpretations," American Birding Association, April 11, 2025, www
.aba.org/aba-recording-rules-and-interpretations/.

98   **In her book:** Valarie Kaur, *See No Stranger* (One World, 2021).

99   **The music of a yoik:** Nils-Aslak Valkeapää, "Ett sätt att lugna renar (A Way
to Calm Reindeer)," lecture, *Lásságámmi,* 1984, www.lassagammi.no/ett
-satt-att-lugna-renar.5856901-315504.html.

100 **In his *Sand County Almanac*:** Aldo Leopold. *Sand County Almanac and Sketches Here and There* (Oxford: Oxford University Press, 1972).

101 **"Some people say animals":** Bernie Krause, quoted in Sarah Sweeney, "The Sounds of Nature, As Music," *Harvard Gazette,* October 1, 2012, https://news.harvard.edu/gazette/story/2012/10/the-sounds-of-nature -as-music/.

102 **"the balance between absorptive surfaces":** Garth Paine, "Using the Sounds of Nature to Monitor Environmental Change," *Smithsonian Magazine,* December 28, 2018, www.smithsonianmag.com/science-nature /using-sounds-wild-monitor-environmental-change-180971128/.

103 **"To distinguish between":** Ella Frances Sanders, *Everything, Beautiful: A Guide to Finding Hidden Beauty in the World* (Penguin, 2022).

Chapter 15

104 **The colorful lights:** Holly Large, "Biofluorescence vs Bioluminescence: What's the Difference?" *IFLScience,* February 13, 2025, www.iflscience .com/biofluorescence-vs-bioluminescence-whats-the-difference-78046.

105 **Humans emit their own:** Masaki Kobayashi, Daisuke Kikuchi, and Hitoshi Okamura, "Imaging of Ultraweak Spontaneous Photon Emission from Human Body Displaying Diurnal Rhythm," *PLOS One,* July 16, 2019, https://doi.org/10.1371/journal.pone.0006256.

106 **A Canadian research team:** Issy Ronald, "Orcas Are Bringing Humans Gifts of Food — but Why?" CNN, July 9, 2025, https://edition.cnn .com/2025/07/09/science/orcas-humans-gifts-food-intl-scli.

107 **British researchers at Queen Mary University:** Nick Petrić Howe and Benjamin Thompson, "These Tiny Fish Combine Electric Pulses to Probe the Environment," *Nature Podcast,* March 6, 2024, www.nature.com /articles/d41586-024-00685-6.

108 **Electric pulses:** Jack Tanusuea, "These Electric Fish Detect Images of What Their Companions Are 'Seeing.'" *Scientific American,* March 6, 2004, http://www.scientificamerican.com/article/these-electric-fish-detect -images-of-what-their-companions-are-seeing.

109 **To save lives:** Erik T. Frank et al., "Wound-Dependent Leg Amputations to Combat Infections in an Ant Society," *Current Biology* 34, no. 14 (2024): 3273–78.e3, https://doi.org/10.1016/j.cub.2024.06.021.

110 ***Harvard Magazine* asks:** Nancy Walecki, "Can Slime Molds Think?"

Harvard Magazine, November–December 2021, www.harvardmagazine
.com/2021/10/right-now-can-slime-molds-think.

111 **They can also:** University of Warwick, School of Life Sciences, "Slime Moulds,"
October 23, 2023, https://warwick.ac.uk/fac/sci/lifesci/outreach/slimemold/.

112 **"including everything from microbes":** Mark A. Anthony, S.
Franz Bender, and Marcel G. A. van der Heijden, "Enumerating Soil
Biodiversity," *PNAS* 120, no. 33 (August 7, 2023), https://doi.org/10.1073
/pnas.2304663120.

113 **Murray told CNN:** Nell Lewis, "Extraordinary Photos Reveal the Secret
Kingdom of the Soil," *CNN Travel,* November 8, 2023, www.cnn.com
/travel/extraordinary-soil-animals-andy-murray-photography-c2e-spc
-scn-intl/index.html.

114 **In November 2023:** Lewis, "Extraordinary Photos."

115 **Consider, too, the phenomenon:** Ferris Jabr, *Becoming Earth: How Our
Planet Came to Life* (Random House, 2024).

116 **"Contrary to long-standing maxims":** Ferris Jabr, "Earth's Mysterious,
Deep-Dwelling Microbes We're Only Starting to Understand," *New York
Times,* June 24, 2024, www.nytimes.com/2024/06/24/magazine/earth
-geomicrobiology-microbes.html.

117 **In his seminal book:** William James, *The Principles of Psychology,* vol. 1
(Henry Holt and Company, 2018).

118 **According to Weil's biographer:** Robert Zaretsky, *The Subversive Simone
Weil: A Life in Five Ideas* (University of Chicago Press, 2023).

119 **Among the most eloquent:** Peter Wohlleben, *The Hidden Life of Trees:
What They Feel, How They Communicate — Discoveries from a Secret
World* (Greystone Books, 2016).

120 **He believes:** Jacob White and Felipe Yamashita, *"Boquila Trifoliolata*
Mimics Leaves of an Artificial Plastic Host Plant," *Plant Signaling &
Behavior* 17, no. 1 (2021), https://doi.org/10.1080/15592324.2021.1977530.

121 **Meanwhile, according to the BBC:** Pallab Ghos, "Animals React to Secret
Sounds from Plants, Say Scientists," BBC, July 14, 2025. https://www.bbc
.com/news/articles/c8e4860n9rpo

122 **This is the same team:** Itzhak Khait et al., "Sounds Emitted by Plants
Under Stress Are Airborne and Informative," *Cell* 186, no. 7 (2023):
1328–36.e10, https://doi.org/10.1016/j.cell.2023.03.009.

123 **The director of the Minimal Intelligence Lab:** Amanda Gefter, "What Plants Are Saying About Us," *Nautilus,* March 7, 2023, https://nautil.us /what-plants-are-saying-about-us-264593/.

124 **"Plants reverberate":** Henderson, Caspar Henderson, "What Plants Hear," *Nautilus,* February 6, 2024, https://nautil.us/what-plants-hear-509994.

125 **Multiple scholars:** "Debunking a myth: plant consciousness." PubMed, National Library of Medicine, November 16, 2020, https://pubmed.ncbi .nlm.nih.gov/33196907/.

126 **"Plants are self-organizing":** Gefter, "What Plants Are Saying about Us."

127 **A year later:** Matt Hayes, "CROPPS Hackathon Ushers in New Era of Plant Communication," *Cornell Chronicle,* April 28, 2025, https://news .cornell.edu/stories/2025/04/cropps-hackathon-ushers-new-era-plant -communication.

128 **The practical goal:** Center for Research on Programmable Plant Systems (CROPPS), "Opening a Dialogue with Plants," https://cropps.cornell.edu.

Chapter 16

129 **"Myths are like palimpsests":** Sophie Strand, *The Body Is a Doorway: A Journey Beyond Healing, Hope and the Human* (Running Press Adult, 2025)

130 **To Strand:** "Rerooting Mythology And Ecological Storytelling: A conversation with Sophie Strand about myth-making and science," *Advaya,* https://advaya.life/magazine/read/rerooting-mythology-and -ecological-storytelling.

131 **In 2017, Hilton Japyassú:** Hilton F. Japyassú and Kevin N. Laland, "Extended Spider Cognition," *Animal Cognition* 20, 3 (2017): 375–95, https://doi.org/10.1007/s10071-017-1069-7.

132 **In 2020, *New Scientist*:** David Robson, "Spiders Think with Their Webs, Challenging Our Ideas of Intelligence," *New Scientist,* February 5, 2020, www.newscientist.com/article/mg24532680-900-spiders-think-with-their -webs-challenging-our-ideas-of-intelligence/.

133 **In Miles's words:** Ron Miles, "Spiders Use Webs to Extend Their Hearing," *ScienceDaily,* 2022, www.sciencedaily.com/releases/2022/03/220329114714.htm.

134 **That same year:** Denyse O'Leary, "Spiders Are Smart; Be Glad They Are Small," *Mind Matters,* December 25, 2022, https://mindmatters.ai/2022/12 /spiders-are-smart-be-glad-they-are-small-2.

135 **Alexander von Humboldt:** Andrea Wulf, *The Invention of Nature: Alexander von Humbolt's New World* (Vintage Books, 2015).

136 **"Western philosophy is preoccupied":** Thích Nhất Hạnh, *The Other Shore: A New Translation of the Heart Sutra with Commentaries* (Paralax, 2017). (Emphasis in original.)

137 **While animals communicate:** Richard Louv, *Our Wild Calling* (Algonquin Books, 2020).

138 **In 2000, the prestigious journal:** "Mysterianism Lite," *Nature Neuroscience* 3, no. 3 (2000): 199, https://doi.org/10.1038/72893.

139 **Writing on *Medium*:** David A. Palmer, "Enchanting the World of Nature," *Medium,* March 20, 2021, https://medium.com/the-new-mindscape /enchanting-the-world-of-nature-3748d2d70f32.

140 **"We need not look far":** Tina Fields, "Trees in Early Irish Law and Lore: Respect for Other-Than-Human Life in Europe's History," *Ecopsychology* 12, 2 (June 2020), https://indigenize.wordpress.com/wp-content /uploads/2020/06/fields-irish-tree-paper-pub-eco.2019.0058.pdf.

141 **Palmer writes:** Palmer, "Enchanting the World of Nature."

142 **This explanation:** Harold Coward, "Hindu Views of Nature and the Environment," *Science Across Cultures: The History of Non-Western Science* 4 (2003): 411–19, https://doi.org/10.1007/978-94-017-0149-5_21.

143 **"The Inuk word *sila*":** Johnny Langenheim, "Natural Custodians: Indigenous Lessons in Reconnecting with Nature," *National Geographic,* January 25, 2024, www.nationalgeographic.com/environment/article /paid-content-natural-custodians-Indigenous-lessons-in-reconnecting -with-nature.

Chapter 17

144 **"The Kumeyaay created":** Angelina Lutz, "Land of Extremes: Kumeyaay Cosmology," *Imperial Valley Press* (El Centro, CA), August 11, 2023, www.ivpressonline.com/life/desertmuseum/land-of-extremes -kumeyaay-cosmology/article_9d7d1000-37c1-11ee-93e3-37060f16d9f8 .html.

145 **The Dutch researchers:** Agnes Van den berg and Cecil Konijnendijk, "Ambivalence Towards Nature and Natural Landscapes," *Environmental Psychology* (BMS Blackwell), January 2012.

146 **Today, one can see:** A *New Yorker* video shows Theo Jansen's wind
-powered Strandbeasts in action, https://www.youtube.com/watch?v=Pj
-NqWDH2qE.

147 *Biophilia*: Edward O. Wilson, *Biophilia* (Harvard University Press), 1984.

148 **University of California:** Michael Saler, "Modernity and Enchantment: A
Historiographic Review," *History Cooperative,* https://historycooperative
.org/journal/modernity-and-enchantment-a-historiographic-review/.

149 **The document describes:** "The New York Declaration on Animal
Consciousness," New York University, April 19, 2024, https://sites.google
.com/nyu.edu/nydeclaration/declaration.

150 **In an article:** John Favini, "What If Competition Isn't as 'Natural' as We
Think?" *Slate,* January 23, 2020, https://slate.com/technology/2020/01
/darwin-competition-collaboration-evolutionary-biology-climate
-change.html.

Chapter 19

151 **In 2023, South African researchers:** Liana Zanette et al., "Fear of
the Human 'Super Predator' Pervades the South African Savanna,"
Current Biology 33, 21 (2023), https://doi.org/10.1016/j.cub.2023
.08.089.

152 **According to the Franklin Institute:** Franklin Institute, "Did Benjamin
Franklin Want the National Bird to Be a Turkey?" Franklin Institute,
Science and Education, https://fi.edu/en/science-and-education/benjamin
-franklin/national-bird.

153 **By 2022, worldwide:** National Park Service, "World CA Condor
Update: 2022 Population Status," www.nps.gov/articles/000/caco
-world-2022.htm.

154 **In 2017, *Sun* magazine:** Al Kesselheim, "A Walk on the Wild Side:
Joe Hutto's Life with Animals," *Sun Magazine,* May 2017, www
.thesunmagazine.org/issues/497/a-walk-on-the-wild-side.

Chapter 20

155 **Not far from here:** *Kumeyaay Sense of the Land and Landscape,*
Viejas Band of Kumeyaay Indians website. October 18, 2016, https://
viejasbandofkumeyaay.org/viejas-community/kumeyaay-history
/kumeyaay-sense-of-the-land.

156 **"Tracking is akin":** Jessica Carew Kraft, *Why We Need to Be Wild: One Woman's Quest for Ancient Human Answers to 21st Century Problems* (Sourcebooks, 2023).

157 **"One of the countless signs":** David Abram, *Becoming Animal: An Earthly Cosmology* (Pantheon Books, 2010).

158 **Beginning in the Paleolithic:** Kumeyaay Information Village and Website, "Kumeyaay History: 12,000 Years, 600 Generations in San Diego," www .kumeyaay.info/history/.

Chapter 23

159 **To deal with despair:** Anthony Paul Smith, "Reflections on Melancholia and Ecology: An Introduction," *Environmental Critique* (blog), April 22, 2012, https://environmentalcritique.wordpress.com/2012/04/22/reflections -on-melancholia-and-ecology-an-introduction/.

160 **In his journal:** Angie Brenner, "Seeking Beauty, Jim's Gift," Tortoise Film & Publishing Company (2021).

Chapter 24

161 **The great columnist Ernie Pyle:** Ernie Pyle, *At Home with Ernie Pyle* (Indiana University Press, 2016).

162 **The Canadian scientist:** Emmanuel Vaughan-Lee, "Finding the Mother Tree — with Suzanne Simard," *Emergence,* October 26, 2022, https:// emergencemagazine.org/conversation/finding-the-mother-tree/.

Chapter 25

163 **Her story:** Kathleen Lockyer, *Wild Inside: How Nature Protects Your Child's Mental Health and Restores Yours* (Otter Play Publishing, 2025).

164 **"Symptoms associated":** Judy Wu, Gaelen Snell, and Hasina Samji, "Climate Anxiety in Young People: A Call to Action," *Lancet* 4, 10 (2020), www.thelancet.com/journals/lanplh/article/PIIS2542 -5196(20)30223-0/fulltext.

165 **Among other problems:** American Psychological Association, Climate for Health, and ecoAmerica, *Mental Health and Our Changing Climate: Impacts, Implications, and Guidance,* March 2017.

166 **The American Psychiatric Association:** American Psychiatric Association, "Get Outside: The Many Ways Nature Can Boost Your Mental

Well-Being," *APA Blog,* November 25, 2004, www.psychiatry.org/news
-room/apa-blogs/ways-nature-can-boost-your-mental-wellbeing.

167 **Humans, she pointed out:** Maria Popova, "The Tree in Me: A Tender
Painted Poem About Growing Our Capacity for Joy, Strength, and Love,"
Marginalian, April 17, 2021, www.themarginalian.org/2021/04/16/the-tree
-in-me-corinna-luyken/.

Chapter 26

168   **Zen master Thich:** Jo Confino, "Beyond Environment: Falling Back in
Love with Mother Earth," interview with Thich Nhat Hanh, *Guardian,*
February 20, 2012, video and transcript available at https://plumvillage
.org/about/thich-nhat-hanh/interviews-with-thich-nhat-hanh/the
-guardian-uk-20-february-2012.

169 **"It is suddenly occurring":** Douglas Brinkley, *Windblown World: The
Journals of Jack Kerouac, 1947–1954* (Penguin), 2006.

170 **The new nature movement:** Richard Louv, "The New Nature Movement
Isn't About Going Back to Nature…Children & Nature Network,"
September 2012, https://www.childrenandnature.org/resources/forward
-to-nature-the-new-nature-movement-isnt-about-going-back-to-nature
-but-forward-to-a-nature-rich-civilization/.

171 **Greener cities:** Children & Nature Network, "Research Digest: Biophilic
Design in Schools," *Children & Nature Network Research Digest,*
September 2024, www.childrenandnature.org/resources/research-digest
-biophilic-design/.

172 **The psychological:** Richard Louv, "Outdoors for All: A Nascent Global
Movement Proclaims That Access to Nature Is a Human Right," *Sierra
Magazine,* May 7, 2019, www.sierraclub.org/sierra/2019-3-may-june
/feature/outdoors-for-all-nature-is-a-human-right.

173 **Consider the city of Oakland:** Devon C. Payne-Sturges et al., "Disparities
in Toxic Chemical Exposures and Associated Neurodevelopmental
Outcomes: A Scoping Review and Systematic Evidence Map of the
Epidemiological Literature," *Environmental Health Perspectives* 131, 9
(2023).

174 **Adjusting urban access:** Patrick Barkham, "Human Connection to
Nature Has Declined 60% in 200 Years, Study Finds," *The Guardian,*

August 9, 2025, www.theguardian.com/environment/2025/aug/09/human
-connection-to-nature-has-declined-60-in-200-years-study-finds.

175   **To slow climate:** Ross Andersen, "Nature Has Lost Its Meaning," *The Atlantic,* November 30, 2015, www.theatlantic.com/science /archive/2015/11/nature-has-lost-its-meaning/417918/161credit162.

176   **I find comfort:** Wendy Webber, "Turning Hope into Action," *State of Formation,* October 21, 2018, https://stateofformation.org/2018/10/turning -hope-into-action/.

Chapter 27

177   **When I walk:** Virginia Woolf, *Moments of Being: Unpublished Autobiographical Writings,* ed. Jeanne Schulkind (Harcourt Brace Jovanovich, 1976).

178   **Since 2008, the beetle:** Vanessa Lopez et al., "The Goldspotted Oak Borer," Center for Invasive Species Research, https:/cisr.ucr.edu/invasive-species /goldspotted-oak-borer.

ABOUT THE AUTHOR

RICHARD LOUV IS A JOURNALIST and the author of eleven books, including the *New York Times* bestseller *Last Child in the Woods*, which helped launch an international movement to connect children, families, and communities to the natural world. He is cofounder of the nonprofit Children & Nature Network, and has written for the *New York Times, Outside, Orion, Parents, Psychology Today,* many other publications. He appears regularly on national radio and TV, and lectures throughout the world. In 2008, he was awarded the Audubon Medal. Prior recipients have included Rachel Carson, E. O. Wilson, Jimmy Carter, and Sir David Attenborough.

RAISING READERS
Books Build Bright Futures

Thank you for reading this book and for being a reader of books in general. We are so grateful to share being part of a community of readers with you, and we hope you will join us in passing our love of books on to the next generation of readers.

Did you know that reading for enjoyment is the single biggest predictor of a child's future happiness and success?

More than family circumstances, parents' educational background, or income, reading impacts a child's future academic performance, emotional well-being, communication skills, economic security, ambition, and happiness.

Studies show that kids reading for enjoyment in the US is in rapid decline:

- In 2012, 53% of 9-year-olds read almost every day. Just 10 years later, in 2022, the number had fallen to 39%.
- In 2012, 27% of 13-year-olds read for fun daily. By 2023, that number was just 14%.

Together, we can commit to **Raising Readers** and change this trend. How?

- Read to children in your life daily.
- Model reading as a fun activity.
- Reduce screen time.
- Start a family, school, or community book club.
- Visit bookstores and libraries regularly.
- Listen to audiobooks.
- Read the book before you see the movie.
- Encourage your child to read aloud to a pet or stuffed animal.
- Give books as gifts.
- Donate books to families and communities in need.

BOB1217

Books build bright futures, and **Raising Readers** is our shared responsibility.

For more information, visit **JoinRaisingReaders.com**

Sources: National Endowment for the Arts, National Assessment of Educational Progress, WorldBookDay.com, Nielsen BookData's 2023 "Understanding the Children's Book Consumer"